Multiple Detection in Size-Exclusion Chromatography

ACS SYMPOSIUM SERIES **893**

Multiple Detection in Size-Exclusion Chromatography

Andre M. Striegel, Editor

The Florida State University.

Sponsored by the
**ACS Divisions of Analytical Chemistry
Polymer Chemistry, Inc.
Polymeric Materials: Science and Engineering, Inc.**

American Chemical Society, Washington, DC

Library of Congress Cataloging-in-Publication Data

Symposium "Size-Exclusion Chromatography with Multiple Detection Techniques" (2003 ; New Orleans, La.)

Multiple Detection in Size-Exclusion Chromatograqphy / André M. Striegel, editor ; sponsored by the ACS Divisions of Analytical Chemistry, Polymer Chemistry, Inc., [and] Polymeric Materials: Science and Engineering, Inc.

p. cm.—(ACS symposium series ; 893)

"Developed from ... the 225th National Meeting of the American Chemical Society, New Orleans, Louisiana, March 23–27, 2003"—Front matter.

Includes bibliographical references and index.

ISBN 0–8412–3878–2 (alk. paper)

1. Gel permeation chromatography—Congresses.

I. Striegel, André M., 1967- II. American Chemical Society. Division of Analytical Chemistry. III. American Chemical Society. Division of Polymer Chemistry, Inc. IV. American Chemical Society. Division of Polymeric Materials: Science and Engineering, Inc. V. American Chemical Society. Meeting (225th : 2003 : New Orleans, La.) VI. Title. VII. Series.

QD272.C444S93 2003
543′.84—dc22   2004053128

PRINTED IN THE UNITED STATES OF AMERICA

Foreword

The ACS Symposium Series was first published in 1974 to provide a mechanism for publishing symposia quickly in book form. The purpose of the series is to publish timely, comprehensive books developed from ACS sponsored symposia based on current scientific research. Occasionally, books are developed from symposia sponsored by other organizations when the topic is of keen interest to the chemistry audience.

Before agreeing to publish a book, the proposed table of contents is reviewed for appropriate and comprehensive coverage and for interest to the audience. Some papers may be excluded to better focus the book; others may be added to provide comprehensiveness. When appropriate, overview or introductory chapters are added. Drafts of chapters are peer-reviewed prior to final acceptance or rejection, and manuscripts are prepared in camera-ready format.

As a rule, only original research papers and original review papers are included in the volumes. Verbatim reproductions of previously published papers are not accepted.

ACS Books Department

Contents

Light Scattering and Viscometry

Fluorescence and UV Absorption Spectroscopy

Mass Spectrometry and NMR Spectroscopy

Chemical and Compositional Heterogeneity and Chemical Composition Distribution

Dynamic Surface-Tension Detection, Data Reduction, and Band Broadening

Indexes

Preface

During the past four decades, size-exclusion chromatography (SEC) has assumed a preeminent role among polymer characterization techniques. It is virtually inconceivable to synthesize a polymer in the laboratory or to study a naturally occurring macromolecule without at least determining its molar mass averages, even if these values are not absolute but merely relative to some calibration standard.

As our understanding of structure–property relationships becomes more fully developed, our synthetic skills more refined, and as we continue to uncover Nature's complexity and macromolecular hierarchies, a number of characterization needs have arisen. We have found a need to determine the absolute molar mass averages and distribution of polymers; to characterize their long- and short-chain branching; conformation; chemical composition, functionality, and heterogeneity; tacticity; copolymer and base-pair sequences; polyelectrolyte charge; and so on. Moreover, in most of these cases, we have also found the need to characterize those properties as a continuous function of the molar mass of the analyte. The ability to perform these measurements is imparted by the multiplicity of detection methods currently in use with SEC: light scattering; viscometry; mass spectrometry; IR, UV/visible, nuclear magnetic resonance, and fluorescence spectroscopy; conductivity; novel methods such as dynamic surface tension; and also by the critical role SEC plays in two-dimensional liquid chromatographic (2D-LC) measurements, where it complements both traditional and cutting-edge separations.

Highlighting the role of the detection methods, and the synergistic effect of combining a variety of methods, was the purpose of a symposium entitled *Size-Exclusion Chromatography with Multiple Detection Techniques,* held at the Spring 2003 American Chemical Society (ACS) National Meeting in New Orleans, Louisiana, which was sponsored by the ACS Divisions of Analytical Chemistry, Polymer Chemistry, Inc., and Polymeric Materials: Science and Engineering, Inc.

The present volume is an outgrowth of that symposium. A number of techniques not represented at the meeting have been included here, such as fluorescence spectroscopy, dynamic surface tension, depolarized light scattering, NMR, inductively coupled plasma- and matrix-assisted laser desorption ionization–MS, and 2D-LC, in addition to a wider variety of applications. It is my hope that this book will provide a detailed overview of the large number of detection methods currently used in conjunction with size-exclusion chromatography, as well as of particular applications and of the determination of fundamental and end-use properties these measurements allow.

In organizing the original Symposium, I distinctly remember calling my colleague Patricia Cotts to invite her to present. I informed her of those speakers who had already agreed to give talks and asked Pat if she would like to join us, to which she enthusiastically replied "Sounds like fun!" This attitude is, of course, characteristic of Pat. During the past year, I have found that it is also characteristic of all those who have contributed to this book. The authors set the highest standards for themselves; kept in close communication about content, format, exposition, and unnecessary duplication of information; and were, without exception, an editor's "Dream Team" with which to work. I am indebted to them for their professionalism, as well as for my own continuing scientific education in the field.

Beyond the authors, I would first like to acknowledge the reviewers, who selflessly gave of their time and expertise and who perforce must labor in anonymity. Their attitude and standards mimicked that of the contributors at all stages. I would also like to thank the ACS Books Department team Bob Hauserman and Stacy VanDerwall in acquisitions and Margaret Brown in editing and production, in particular Stacy VanDerWall, for her quick and thorough responses to all my queries in the acquisitions phase of the book. My thanks also to John Dorsey; to Victoria McGuffin who, as Chair of the Division of Analytical Chemistry's Chromatography and Separations Chemistry subdivision, invited me to chair the original Symposium; and, foremost, to my colleague and friend David Alward who provided extensive personal and professional support at all stages of this project.

André M. Striegel
Department of Chemistry and Biochemistry
The Florida State University
Tallahassee, FL 32306–4390
striegel@chem.fsu.edu (email)

Light Scattering and Viscometry

Chapter 1

Separation Science of Macromolecules: What Is the Role of Multidetector Size-Exclusion Chromatography?

André M. Striegel

Department of Chemistry and Biochemistry, The Florida State University, Tallahassee, FL 32306–4390

The importance of separation science in the study of natural and synthetic macromolecules is described, highlighting the role of size-exclusion chromatography (SEC) and emphasizing that of multiple detection in this analytical technique. Particular attention is paid to the ability of separation science in general, and of multi-detector SEC in particular, to determine the distributions of key macromolecular parameters and the end-use properties these affect. The historical development of SEC is reviewed and mention is made of how non-separation techniques can complement the type of information provided by SEC and other macromolecular separation method.

The end-use application of most macromolecules is determined not only by their chemical identity but also, and sometimes more importantly, by the distributions and sequences of key physical and physicochemical parameters. The distribution of molar mass is known to affect a large number of properties, as for example with elastomers, where narrowing the molar mass distribution (MMD) results in superior mechanical properties but relatively poor processing characteristics. The distribution of properties such as long-chain branching (LCB) and three-dimensional structure and/or conformation (referred to herein collectively as "architecture") can affect areas as dissimilar as the abilities to form inclusion complexes and to modify flow through drilling pipes. Chemical composition distribution can affect the miscibility and morphology of polymer blends, particle size distribution affects the packing and rheological behavior of resins and melts, while charge distribution influences the drug-delivering ability of polymeric sequestering/transfer agents. It is well known that children who inherit sickle cell anemia from both parents rarely live beyond the age of two. This genetic disease is known to arise from a change in ~0.3% of the amino acid sequence of hemoglobin, a change of merely two amino acids out of 574. Figure 1 shows examples of various polymeric distributions; a more comprehensive list is given in Table I along with the types of end-use properties affected.

The ability to determine these and many other distributions is gained through separation science. As seen in Table I, the use of chromatographic and other methods permits the study of a large and varied array of macromolecular property distributions and sequences. Some methods, such as temperature rising elution fractionation (TREF), are intended for use with crystalline polymers such as polyethylene. Rheology, enzymology, and matrix assisted laser desorption/ionization mass spectrometry (MALDI-MS) have been included as well; while not separation methods *per se,* the information they yield is often complementary to that obtained by techniques such as size-exclusion chromatography (SEC). The preeminent role of SEC, in particular of multi-detector SEC techniques, is seen by its abundant representation in the table, which has been highlighted using **boldface** type. The role of the complementary, non-separation methods will be discussed briefly at the end of this chapter.

It has been said that "[t]he only completely satisfactory description of the molecular weight (i.e., the degree of polymerization) of a macromolecular compound is the distribution curve...as determined through fractionation."[1] Analogous truisms apply to all the other properties mentioned in Table I. As seen, to achieve the "completely satisfactory description[s]" requires enlisting the aid of a large number of separation methods (many still in their infancy), usually in combination with spectroscopic, hydrodynamic, etc. methods of detection. Though the separation techniques each possess their own individual thermodynamic and kinetic (and instrumental) identities, they are also fundamentally united in ways that allow for their complementarity, study, and improvement.[2] The polymer molecules (including natural and synthetic polymers as well as oligomers) thus play a dual role, that of analyte and that of probe. In the former, separation science is used to enhance our knowledge of the

4

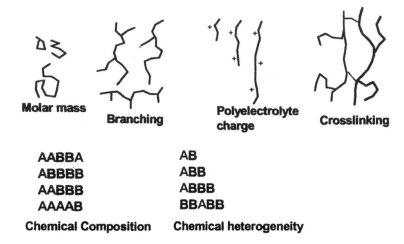

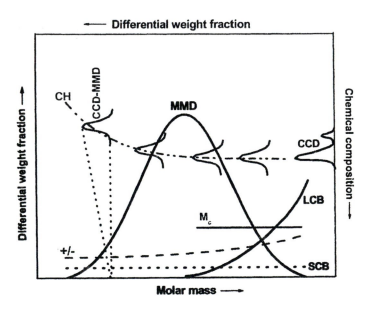

Figure 1. Generic examples of (top) various macromolecular heterogeneities and (bottom) their distributions as a continuous function of molar mass. MMD: molar mass distribution, CCD: chemical composition distribution, CCD-MMD: CCD as a function of MM (this is normally represented as a contour plot), CH: chemical heterogeneity, LCB: long-chain branching, SCB: short-chain branching, M_c: molar mass between crosslinks, +/-: polyelectrolytic charge.

Table I. Macromolecular distributions:
Their measurement and end-use effects

Macromolecular Property	Representative end-use properties affected	Separation method used for determination[a]
Molar mass	Elongation, tensile strength, adhesion	**SEC, FFF, HDC, TGIC, CEC** SFC, MALDI-MS, rheology
Long-chain branching	Shear strength, tack, peel, crystallinity	**SEC-MALS, SEC-VISC,** rheology, enzymology
Short-chain branching	Haze, stress-crack resistance, crystallinity	**SEC-IR, SEC-NMR, TREF[b],** CRYSTAF[b], enzymology
Crosslinking	Gelation, vulcanization, surface roughness	**SEC-MALS, SEC-VISC,** rheology
Architecture	Flow modification, diffusion encapsulation	**SEC-MALS-QELS-VISC**
Tacticity	Crystallinity, anisotropy, solubility	**SEC-NMR,** TGIC, LCCC,
Chemical composition	Morphology, miscibility, solubility	GPEC, TGIC
Chemical heterogeneity	Toughness, brittleness, biodegradability	**SEC-spectroscopy/spectrometry,** LCCC, PFC
Chemical comp. vs. molar mass	Mechanical properties, blending, plasticization	2D-LC (e.g., **SEC-GPEC**)
Block sequence	Dielectric properties, reactivity, miscibility	**SEC-spectroscopy,** 2D-LC (e.g., **PFC-SEC**)
Base pair sequence	Genetic code, heredity, mutations	Automated DNA sequencing, MALDI-MS
Polyelectrolytic charge	Flocculation, transport, binding of metals	**SEC-conductivity**
Particle size	Packing, drag, friction, mixing	FFF, HDC, PSDA, sieving

Many techniques require a concentration-sensitive detector (e.g., a differential refractometer), not included here for simplicity.
[a]SEC: size-exclusion chromatography, FFF: field-flow fractionation, HDC: hydrodynamic chromatography, TGIC: temperature-gradient interaction chromatography, CEC: capillary electrokinetic chromatography, SFC: supercritical fluid chromatography, MALDI-MS: matrix-assisted laser desorption/ionization mass spectrometry, MALS: multi-angle light scattering, VISC: viscometry, IR: infrared spectroscopy, NMR: nuclear magnetic resonance spectroscopy, TREF: temperature rising elution fractionation, CRYSTAF: crystallization fractionation, QELS: quasi-elastic (dynamic) light scattering, LCCC: liquid chromatography at the critical condition, GPEC: gradient polymer elution chromatography, PFC: phase fluctuation chromatography, 2D-LC: two-dimensional liquid chromatography, PSDA: particle size distribution analyzer.
[b]For crystalline polymers only.

macromolecules and to improve, tailor, and predict their end-use properties. In the latter, the analytes themselves help to develop a clearer picture of the fundamental processes, of the commonalities and distinction lying below the alphabet soup that is the "Separation Method" column of Table I.

As our focus is principally on size-exclusion chromatography, we now take a moment to remember how we got to where are, i.e., by briefly reviewing the development of SEC.

Historical Development of SEC

It has been over a half century since Wheaton and Bauman noted the fractionation of non-ionic substances in the passage through an ion exchange column, indicating that the separation of molecules based on size should be possible in aqueous solution.[3] In 1959, Porath and Flodin demonstrated that columns packed with crosslinked polydextran gel, swollen in aqueous media, could be used to separate various water-soluble macromolecules by size.[4] This became known as gel filtration chromatography (GFC). Soon other hydrophilic gels were developed for separation of compounds of biological interest. Being capable of swelling only in aqueous media, however, limited their use to water-soluble substances.

By the early 1960s, the relationship between the molar mass distribution of synthetic polymers and their physical characteristics such as chemical resistance, toughness, melt viscosity, etc. (see Table I) was well known. Consequently, workers in the polymers and plastics fields were interested in a method to obtain not only molar mass averages but, more importantly, molar mass distributions of synthetic polymers.[5] During this time, work had begun on making hydrophobic gels and columns were packed consisting mainly of crosslinked polystyrene, with the crosslinking performed in the absence of diluents. It was soon recognized that crosslinking in the presence of diluents that are solvents for the monomer altered the structure of the gel networks. When the diluent is a non-solvent for the resulting polymer a rugged, stable internal gel structure of good permeability may be obtained.

In 1964, John Moore published the first paper on a technique that he termed gel permeation chromatography (GPC).[6] In said paper, he described the preparation of polystyrene beads of sufficient crosslinking to impart a desired rigidity to the gel while still regulating the permeability by altering the amount and nature of the diluent. The composition of the gels consisted mainly of varying proportions of styrene/divinylbenzene/toluene. Large changes in the permeability of the gels were effected by changing the diluent. Columns packed with these gels were used to separate a series of polystyrenes and of poly(propylene glycol)s over an extended range of molar mass. To monitor the composition of the eluent a differential refractometer was used. This last was a special design, by James Waters, with a smaller optical cell than was commercially available at the time and with provision for continuous flow in both sides of the cell.[5] Moore recognized that the chromatographic separation appeared to be close to an equilibrium process in which the solute molecules diffused rapidly into all available parts of the gel network.[6] The thermodynamic equilibrium of the separation process remains a topic of interest to this day for size-exclusion chromatography (SEC),[7] the all-encompassing term that is used nowadays for both GPC and GFC.[8]

Separation Science and Biopolymers

The field of biopolymers and copolymers is one rich with possibilities and areas of interest. Biodegradable copolymers, in particular, combine biological and synthetic, linear and branched, neutral and polyionic and, occasionally, lightly crosslinked materials for applications in drug delivery,[9] "smart" sutures,[10] and tissue regeneration,[11] for example. Characterizing these copolymers' properties leads directly to improved capabilities and predictive behavior. For example, the biocompatibility and biodegradability of implantation materials made from poly(DL-lactic acid/glycine) copolymers are directly related to both molar mass and chemical composition.[12] Determining the distributions of these and other key functional parameters cannot be accomplished without the use of SEC and related techniques.

Equally interesting, and oftentimes more challenging, is the field of glycopolymers.[13] At the oligomeric level, oligosaccharides perform a large number of biological roles, from nutrition to being essential components of plant cell walls, to moderating biosynthesis, structure, and transport functions of glycoproteins.[14] While there are not many distributions of properties at this level, oligosaccharide analysis is aided, many times critically, by separation techniques such as high-performance liquid chromatography (HPLC) and anion-exchange chromatography (AEC), or by SEC for identification and measurement of key conformational properties.[15]

At the macromolecular level, many botanical glycopolymers possess limited solubility (cellulose); ultra-high molar masses and broad molar mass distributions (amylopectin); long- and short-chain branching (dextran) and, in some cases, hyperbranching (certain Type II arabinogalactans); a variety of anomeric configurations and glycosidic linkages (xanthan, which is also a polyelectrolyte); etc. Dextran sulphate, for example, is a glycopolymer known for its anti-coagulant and bioinhibitory properties (e.g., inhibiting enzyme release from macrophages). It is also a high molar mass, polydisperse, polyanionic, potentially branched polymer. Their use in biological processes and pharmaceutical formulations, adhesives and rheological modifiers, foods and feeds, textiles and non-wovens and, more recently, biodegradable products shows the criticality of applying separation science to the study of these macromolecules. Multi-detector SEC can determine virtually all of the properties mentioned in this section, exceptions being the chemical composition distribution, determined by *e.g.*, gradient polymer elution chromatography (GPEC, though this technique has seen limited application in the study of biopolymers), and the various anomeric configurations and glycosidic linkages, for which we must rely on the help of enzymology and/or mass spectrometry.

Separation Science and Synthetic Polymers

There are a great many synthetic polymers that also possess distributions in a number of parameters critical to their end-use applications. Even a cursory review of the literature, or of this book, will reveal a multitude of examples. In contrast to natural polymers, some control may be exerted over properties such as molar mass and its distribution, branching, copolymerization, etc. in synthetic polymers. Nonetheless, polydispersity may exist not only in molar mass, but also in properties such as tacticity, crystallinity, branching, chemical composition, functionality type, etc. The molar mass distribution becomes but a minimum datum that needs to be measured for many "real world" polymers. It is interesting to note that certain properties may possess both a polydispersity as well as a non-uniform distribution across the MMD. One example, in the case of random copolymers or terpolymers, *e.g.,* a random AB copolymer, is that the ratio of A to B may not be distributed uniformly across the MMD of AB (assuming AB is polydisperse with respect to molar mass). This is referred to as chemical heterogeneity and may be measured by SEC using spectroscopic (IR, NMR) or spectrometric (MS) methods of detection. The spectroscopic detection methods are discussed in the chapters by DesLauriers and by Montaudo, while mass spectrometric detection methods are discussed in the chapters by Montaudo, by Sadi *et al.*, by Lecchi and Abramson, and by Prokai *et al.* The chemical heterogeneity may also be measured using other separation techniques such as liquid chromatography at the critical condition (LCCC, discussed in the chapter by Pasch) or phase fluctuation chromatography (PFC, discussed in the chapter by Teraoka). Regardless of whether the A:B ratio remains constant or not across the MMD of AB, however, there may also be a polydispersity in the amount (mole or weight percent) of either A or B (or both). This is referred to as the chemical composition distribution (CCD) and, as mentioned above, must be determined using non-SEC methods such as GPEC[16,17] or temperature-gradient interaction chromatography (TGIC).[18] Generic examples of chemical heterogeneity and chemical composition distribution are shown in Figure 1. As seen in Table 1, the chemical heterogeneity and the chemical composition distribution can each influence different end-use properties of materials; thus, both parameters need to be controlled and measured accurately for copolymers.

Another example is that of short-chain branching (SCB). As described in the chapter by DesLauriers, SEC-FTIR may be used to measure the distribution of short-chain branches across the MMD of polyolefins. Theoretically, at least, this should also be possible by SEC with ^{13}C-NMR detection, though this author has been unable to find any reference to this in the literature; however, SEC-NMR is still in its infancy though, as described in the chapter by Montaudo, it is quickly coming of age. Additionally, the SCB may also have a distribution and a polydispersity, and recently this has been measured using analytical temperature rising elution fractionation (A-TREF).[19]

One advantage in studying synthetic polymers is the ability to make materials such that individual parameters can be isolated. Polyethylene, for example, has been a benchmark polymer in the study of long-chain branching (LCB), as it fulfills all the requirements that are necessary for accurate, quantitative calculation of LCB *via* the classic Zimm-Stockmayer theory:[20,21] Linear standards exists with the same chemistry as the branched material; the standards cover the MMD region of interest of the branched material; the branching functionality is usually known *a priori* due to a refined understanding of free-radical, Ziegler-Natta, etc. polymerization mechanisms; and there are even a modest amount of relatively narrow molar mass polydispersity standards commercially available for this polymer. The ability to isolate parameters to study their individual effect(s) is also seen in polystyrene, where narrow molar mass polydispersity standards exist over several orders of magnitude in molar mass, and where linear, broad molar mass polydispersity PS may be compared to *f*-functional stars with varying, but controlled, number of arms.[22] Many synthetic polymers do not possess these conveniences, however, and in these cases (in the absence of superior synthetic strategies) assumptions and extrapolations need to be made in order to advance our knowledge of the field. These "real world" polymers are abundant and many possess a number of the distributions given in Table I. An example, studied by this author and others, is poly(vinyl butyral) or PVB, the main component of the polymeric interlayer in laminated safety glass.[23] This macromolecule is actually a random terpolymer, with polydispersities in molar mass (as shown by SEC-MALS)[23] and chemical composition (determined by GPEC),[16] perhaps possessing chemical heterogeneity (though one study has shown otherwise, using SEC-IR),[24] with long- (shown by SEC-MALS-VISC)[23] and possibly short-chain branching and, occasionally, crosslink- or graft-induced branching as well (SEC-MALS-VISC).[23] Additionally, PVB lacks narrow molar mass polydispersity standards or adequate (i.e., same chemical heterogeneity and CCD) linear standards for branching calculations, and possesses intra- and inter-molecular hydrogen-bonding the affects solubility and dissolution. As mentioned, in order to further our knowledge of this and other "real world" polymers assumptions, compromises, and extrapolations must be made, and we should always remember that any of these could be wrong. For more information on MALS and VISC as detection methods in SEC, as well as on QELS and refractometry, the reader is referred to the chapters by Reed and by Cotts. Data handling for refractometry, static light scattering, and viscometry is discussed in the chapter by Brun, and a number of applications of SEC-MALS are given by Podzimek in his chapter.

Polymers as Analytical Probes

While obviously an area rich with possibilities, we will deal here only briefly with the use of polymers as analytical probes, i.e., as used to shed light on the separation processes and methods.

Polymers may be used to determine reduced column factors, self-similarity features of separation media,[25] etc. They may be used to study chromatographic band broadening,[26] as seen in the chapter by Netopilík, as well as in the study of local polydispersity effects that can plague hydrodynamic-volume-dependent separations such as SEC.[27] More generally, polymers can serve to demonstrate the limitations and biases of current techniques, such as the apparent limits of SEC, as compared to TGIC, toward characterizing narrow polydispersity polymers.[28] Macromolecules can also serve to showcase the advantages of new methodologies, such as phase fluctuation chromatography (PFC) for determination of the chemical composition distribution of copolymers. This technique, and its coupling to SEC, is explained in depth in the chapter by Teraoka.

Conclusions

Multi-detector SEC plays a pivotal role in the study of natural and synthetic macromolecules. This technique (or these techniques) has the ability to measure an abundance of parameters and, more importantly, their distributions. It is not, however, the end-all/be-all of analytical methods and is oftentimes aided by other members of the large family of separation methods to which it belongs. Incestuously, it tends to couple with some of these other methods in the form of two-dimensional liquid chromatographic separations.

Multi-detector SEC can also be intimately liked to other, non-separation methods. Liaisons with enzymic and mass spectrometric methods have been hinted at above. The generally tepid relationship between SEC and rheology is slowly warming, as separations scientists recognize the ability of rheology to measure LCB, as well as supra-molecular properties, of materials that may be difficult to analyze chromatographically due to problems with dissolution, non-size-exclusion effects during separation, etc. and rheologists recognize the ability of SEC to measure the distribution of parameters such as

LCB and SCB across the MMD of a polymer (oftentimes with orders of magnitude less sample than necessary for rheology studies), data unavailable from rheological measurements. More often that not these two techniques have the ability to complement each other, and this ability is slowly being exploited by researchers.[29-31]

In the meantime, seemingly unrelated techniques are entering the fray, as with the case of atomic force microscopy (AFM), where a recent report has demonstrated the measurement of the number-average molar mass (M_n) as well as the MMD by combining AFM with the Langmuir-Blodget technique.[32] This is also the case with dynamic surface tension which, as described in the chapter by Synovec *et al.*, is now being used as a detection method in SEC. Witness also the growing number of applications for fluorescence detection in SEC, covered in this book by Maliakal *et al.* and by Yokoyama and Knuckles, where each group applies the technique to completely different ends than the other.

Finally, the determination of polymer size, conformation, etc. can now be aided intimately by the variety of computer modeling methods available, including *ab initio,* semi-empirical, Monte Carlo, molecular mechanics, and molecular dynamics techniques, among others.[33-35]

References

1. Vollmert, B. *Polymer Chemistry;* Springer-Verlag: New York, 1973.
2. Giddings, J. C. *Unified Separation Science;* Wiley-Interscience: New York, 1991.
3. Wheaton, R. M.; Bauman, W. C. *Ann. NY Acad. Sci.* **1953,** *57,* 388.
4. Porath, J.; Flodin, P. *Nature* **1959,** *183,* 1657.
5. Moore, J. C. *J. Polym. Sci.* **1968,** *C-21,* 1.
6. Moore, J. C. *J. Polym. Sci.* **1964,** *A-2,* 835.
7. Striegel, A. M. *J. Chromatogr. A* **2004,** *1033,* 241.
8. Yau, W. W.; Kirkland, J. J.; Bly, D. D. *Modern Size-Exclusion Liquid Chromatography. Practice of Gel Permeation and Gel Filtration Chromatography;* Wiley-Interscience: New York, 1979.
9. Kissel, T.; Jung, T.; Kamm, W.; Breitenbach, A. *Macromol. Symp.* **2001,** *172(Polymers in Medicine),* 113.
10. Lendlein, A.; Langer, R. *Science* **2002,** *296,* 1673.
11. Yannas, I. V. *Adv. Polym. Sci.* **1995,** *122(Biopolymers II),* 219.
12. Schakenraad, J. M.; Dijkstra, P. J. *Clinical Materials* **1991,** *7,* 253.
13. Dwek, R. A. *Biochem. Soc. Trans.* **1995,** *21,* 1.
14. Varki, A. *Glycobiology* **1993,** *3,* 97.
15. Striegel, A. M. *J. Am. Chem. Soc.* **2003,** *125,* 4146 (see Erratum in *J. Am. Chem. Soc.* **2004,** *126,* 4740).
16. Striegel, A. M. *J. Chromatogr. A* **2002,** *971,* 251.

12

17. Striegel, A. M. *J. Chromatogr. A* **2003**, *996*, 45.
18. Chang, T.; Lee, W.; Lee, H. C.; Cho, D.; Park, S. *Am. Lab.* **2002**, *34*, 39.
19. Mirabella, F. M. *J. Polym. Sci. B: Polym. Phys.* **2001**, *39*, 2819.
20. (a) Zimm, B. H.; Stockmayer, W. H. *J. Chem. Phys.* **1949**, *17*, 1301. (b) Striegel, A. M. in *Encyclopedia of Chromatography;* Cazes, J., Ed.; Marcel Dekker: New York, 2001; p. 497.
21. (a) Striegel, A. M.; Krejsa, M. R. *J. Polym. Sci. B: Polym. Phys.* **2000**, *38*, 3120. (b) Striegel, A. M.; Krejsa, M. R. *Int. GPC Symp. Proc.* **2000.**
22. Striegel, A. M. *J. Biochem. Biophys. Meth.* **2003,** *56*, 117.
23. Striegel, A. M. *Polym. Int.* in press.
24. Cotts, P. M.; Ouano, A. C. in *Microdomains in Polymer Solutions;* Dubin, P., Ed.; Plenum: New York, 1985; p. 101.
25. García-Lopera, R.; Irurzun, I.; Abad, C.; Campos, A. *J. Chromatogr. A* **2003,** *996*, 33.
26. Striegel, A. M. *J. Chromatogr. A* **2001,** *932*, 21.
27. Mourey, T. H.; Balke, S. T. *J. App. Polym. Sci.* **1998,** *70*, 831.
28. Lee, W.; Lee, H.; Cha, J.; Chang, T.; Hanley, K. J.; Lodge, T. P. *Macromolecules* **2000,** *33*, 5111.
29. Lusignan, C. P.; Mourey, T. H.; Wilson, J. C.; Colby, R. H. *Phys. Rev. E* **1999,** *60*, 5657.
30. Gillespie, D.; Cable, K.; King, T.; Yau, W. W. *TAPPI J.* **1999,** *82*, 188.
31. Striegel, A. M.; Alward, D. B. *J. Liq. Chromatogr. Rel. Technol.* **2002,** *25*, 2003 (see Erratum *in J. Liq. Chromatogr. Rel. Technol.* **2003,** *26*, 157).
32. Sheiko, S. S.; da Silva, M.; Shirvaniants, D.; LaRue, I.; Prokhorova, S.; Moeller, M.; Beers, K.; Matyjaszewski, K. *J. Am. Chem. Soc.* **2003,** *125*, 6725.
33. Striegel, A. M.; Plattner, R. D.; Willett, J. L. *Anal. Chem.* **1999,** *71*, 978.
34. Freire, J. J. *Adv. Polym. Sci.* **1999,** *143*, 35.
35. Gelin, B. R. *Molecular Modeling of Polymer Structures and Properties;* Hanser: Munich, 1994.

Chapter 2

Fundamentals of Static Light Scattering and Viscometry in Size-Exclusion Chromatography and Related Methods

Wayne F. Reed

Physics Department, Tulane University, New Orleans, LA 70118

Some of the fundamentals of light scattering and viscosity are presented, and their applications as SEC detectors discussed and illustrated. Examples of the types of problems that can be solved by multidetector SEC are also given; phase separation in multicomponent systems, and characteristics of multimodal polymer populations. The power of multidetector SEC to characterize polymers far beyond what is possible by conventional column calibration is highlighted. Finally, some recent innovations that use the same type of coupled multiple detectors, but without SEC columns, are briefly surveyed, which allow for powerful alternative and complementary means of characterizing equilibrium and non-equilibrium properties of polymer solutions.

Light Scattering and Viscosity Background

Light Scattering Notions[*]

The mathematical description of light scattering by molecules was developed by Lord Rayleigh[1] using the then recently formulated Maxwell's equations. Among his major findings was the dependence of scattering intensity on the inverse fourth power of the incident wavelength. This explains why the

---

[*] Only total intensity light scattering, sometimes termed 'static light scattering' , is dealt with here. Dynamic light scattering (DLS) is concerned with time autocorrelation of scattered intensity fluctuations that can be directly related to particle mutual diffusion coefficients and other dynamic effects. Similarly, Raman and Brillouin scattering, in which there is a partial energy transfer to or from internal rotovibrational or acoustic modes are not treated here.

sky is blue, and why at sunrise and sunset the vicinity of the sun is reddish.[2] Later, Einstein used fluctuation theory to describe how pure liquids scatter light, even though perfect crystals scatter none.[3] Subsequently, in the 1940s, Debye[4] and Zimm[5] related the scattering of light to the polarizability, masses, and interactions of polymers in solution. Since then, 'total intensity light scattering' measurements have been arduously applied to characterizing the equilibrium properties of polymers, including copolymers.[6,7] Numerous advances in diode laser sources, fiber optics, ultra-sensitive light detection, and high speed microcomputers have now allowed light scattering to be applied in increasingly powerful ways to solve problems involving both biological and synthetic macromolecules, its application to SEC being one of the most notable recent advances.

Light scattering arises from the interaction of the electric and magnetic fields in the incident light wave with the electron cloud distribution in a scattering particle. The mechanism of interaction involves the induction of electric and magnetic dipoles, quadrupoles, and higher poles in the scatterer, that oscillate at or about the frequency of the incident light. The fundamentals of such scattering are treated in detail in standard texts.[8] In most polymers of interest the electric dipole scattering mechanism is predominant, and that is the type of light scattering treated here. It should be noted that metallic and other conducting particles entail significant magnetic dipole radiation, but are omitted from this chapter. The theory behind light scattering is directly applicable to electromagnetic radiation of any wavelength, and the theory scales as the ratio of the scatterer's dimension to the wavelength of light, hence scattering basics translate easily into other areas such as meteorology, air quality control, and astrophysics.

Almost all modern light scattering detectors (LS) for polymer solution characterization use vertically polarized incident light from a LASER and one or more detectors in the horizontal plane, often termed the 'scattering plane'. For scatterers with scalar polarizability (i.e. the induced dipoles are parallel to the incident electric field) the intensity of scattered light is a maximum in the scattering plane. The detection angle θ in the scattering plane runs from 0^0 for light propagating in the direction of the incident beam, to 180^0 for fully back-scattered light. For particles with non-scalar polarizability, such as rods and ellipsoids, the direction of the induced dipoles is not aligned with the electric field, causing 'depolarized' scattering to occur.

It is instructive to consider the result for a Rayleigh scatterer with scalar polarizability α. Unless otherwise noted cgs units are used throughout this work. A 'Rayleigh' scatterer is a particle whose characteristic linear dimension is much smaller than the wavelength of incident light, such that there is no angular dependence in the scattering plane. The scattered intensity (power/area) at a distance r from the scatterer, and at an altitude angle ϕ (where the scattering plane is at $\phi=90^0$), is represented by $I_s(r,\phi)$, and is given in terms of the incident intensity I_0, the incident wavelength λ and α according to

$$I_s(r,\phi) = I_0 \sin^2 \phi \frac{(2\pi)^4 \alpha^2}{\lambda^4 r^2} \tag{1}$$

The salient features are that the scattering is proportional to i) $1/\lambda^4$, as noted above, ii) α^2, that is, how susceptible the electron cloud of the scatterer is to an applied electric field, and iii) $1/r^2$. Equation 1 shows that for a given incident wavelength and intensity the scattered intensity at a distance r and azimuth angle ϕ depends *only* on the polarizability α; i.e. light scattering is a fundamental interaction between electromagnetic waves and matter and does not involve any arbitrary assumptions, empirical fitting parameters, or statistical models. In this sense it is often asserted that light scattering provides an *absolute* characterization of polymers. The value of α, however, resides in the complicated quantum mechanical nature of the electron distribution in a given scatterer, and, whereas theories exist for its computation, it is easier to measure α via the macroscopic *index of refraction*. This is explained below.

Dividing the scattered intensity per unit volume occupied by scatterers by the incident intensity, and multiplying by r^2 eliminates the dependence on the detectors' distance from the scattering volume, and creates a quantity, the *Rayleigh Scattering Ratio* R (cm^{-1}),[†] which can be interpreted as the fraction of the incident intensity scattered per steradian of solid angle per centimeter of scattering media traversed. R is known to high precision for several pure liquids. For example, $R=1.069 \times 10^{-5}$ (cm^{-1}) for toluene at $T=25^0C$ when light of $\lambda=677nm$ is incident. This means, in practical terms, that R for any polymer solution can be determined simply by comparing the ratio of the scattering detector voltage from the polymer solution to the voltage found by scattering from pure toluene. Hence, a solvent such as toluene firmly anchors light scattering measurements to absolute values of R. In turn R is related to fundamental polymer properties.[‡]

[†] The Rayleigh scattering ratio is often represented as I_R or even simply as I.

[‡] It has apparently become virtually standard practice for SEC/LS practitioners to calibrate R via the scattering from a polymer standard (e.g. low molecular weight polystyrene) instead of using a pure reference solvent such as toluene. This is not generally a good practice, because it requires that M_w of the standard be well known, when, in fact such standards can degrade in time. Aqueous standards have a further risk of being aggregated or displaying other anomalies.

Integrating R over all solid angle gives the total fraction of incident light lost to scattering per cm of pathlength, that is, the turbidity τ. For a Rayleigh scatterer, with vertically polarized incident light

$$\tau = \frac{8\pi}{3} R \qquad (2)$$

The intensity of a propagating beam of light in one dimension diminishes due to turbidity according to

$$I(x) = I_0 e^{-x\tau} \qquad (3)$$

For example, light of $\lambda = 677$nm propagating in toluene at 25^0C will travel 77 meters before its intensity drops to one half its original value.

The power of making absolute determinations of R is apparent when the well known Zimm equation is considered, in which R(c,q) is determined as both a function of scattering vector amplitude q, and polymer concentration c (cm^3/g).[5]

$$\frac{Kc}{R(c,q)} = \frac{1}{MP(q)} + 2A_2c + \left[3A_3Q(q) - 4A_2{}^2MP(q)(1-P(q))\right]c^2 + O(c^3) \qquad (4)$$

where R(c,q) is the excess Rayleigh scattering ratio, that is, the scattering from the polymer solution minus the scattering from the pure solvent, P(q) the particle form factor, A_2 and A_3 are the second and third virial coefficients, respectively, and q is given by

$$q = (4\pi n_0/\lambda)\sin(\theta/2) \qquad (5)$$

K is an optical constant, given for vertically polarized incident light by

$$K = \frac{4\pi^2 n_0{}^2 (\partial n / \partial c)^2}{N_A \lambda^4} \qquad (6)$$

where n_0 is the solvent index of refraction, λ is the vacuum wavelength of the incident light, N_A is Avogadro's number, and Q(q) involves a sum of Fourier transforms of the segment interactions that define A_2.[5] $\partial n / \partial c$ is the differential refractive index for the polymer in the solvent and embodies the Claussius-Mossotti equation for a dilute solution of particle density N, which relates α to the index of refraction n of the polymer solution,

$$n^2 - n_0^2 = 4\pi N\alpha \qquad (7)$$

Most water soluble polymers have a positive value of $\partial n/\partial c$, chiefly because $n_0=1.33$ for water is low compared to most organic substances, whereas organosoluble polymers in organic solvents frequently have negative values of $\partial n/\partial c$. When $\partial n/\partial c = 0$ the polymer/solvent pair is termed 'iso-refractive', and there is no excess scattering due to the polymer. $\partial n/\partial c$ also allows computation of c in equation 4 using a differential refractometer (RI),

$$c = CF\frac{\Delta V_{RI}}{(\partial n/\partial c)} \qquad (8)$$

where ΔV_{RI} is the difference in the RI output voltage between the polymer containing solution and the pure solvent. CF is the calibration factor of the RI (Δn/Volt), and should be periodically checked for accuracy. A convenient means of doing this is by using NaCl solutions, for which the relationship between Δn and [NaCl] is well known at $\lambda=632$nm and T=25^0C. [9]

$$\Delta n=1.766 \times 10^{-3} [NaCl] \qquad (9)$$

where [NaCl] is in grams of NaCl per 100 grams of water. Once CF is known the RI instrument provides an easy means of determining $\partial n/\partial c$ for any polymer/solvent system.

The Zimm equation is the workhorse of light scattering practice in polymer solution analysis and several important polymer characteristics can be determined by measurements of R(c,q); weight average mass M_w, z-average mean square radius of gyration $<S^2>_z$, A_2, A_3, P(q), and Q(q). Extensive literature exists on this topic. Alternative scattering expressions for semi-dilute solutions have also been proposed. [10]

One of the central approximations in the Zimm equation is that the intramolecular interference that leads to the form factor P(q), is based purely on the geometrical path difference that light rays travel from different points on a scattering particle to the detector. This is sometimes called the Rayleigh-Debye approximation [11] and holds as long as

$$\frac{2\pi a\left|\dfrac{n_p}{n_0} - 1\right|}{\lambda} \ll 1 \qquad (10)$$

where n_p is the index of refraction of the particle, and a its characteristic linear dimension. If this condition does not hold, as is likely when *solid* or large dielectric particles scatter light, then Maxwell's equations must be solved, with appropriate boundary conditions, an approach often referred to as 'Mie Scattering'.[11] Fortunately, most polymers are 'threadlike' entities imbued with solvent, so that the main source of optical path difference in scattered rays is indeed the geometrical path difference, and the approximation holds.

Light Scattering Measurements

Equilibrium characterization of polymers usually takes place in dilute solution where $1 >> 2A_2cM_w > > 3A_3c^2M_w$,[§] and over an angular range such that $q^2<S^2> <1$ In this case, the Zimm equation reduces to one of its most frequently used forms:

$$\frac{Kc}{R(c,q)} = \frac{1}{M_w}\left(1+\frac{q^2 <S^2>_z}{3}\right)+2A_2c \qquad (11)$$

A typical batch experiment involves measuring $R(c,q)$ over a series of angles for several polymer solutions at different concentration. If for each concentration the angular data is extrapolated to $q=0$, and these points are then fit, the slope will yield $2A_2$, and the y-intercept will be $1/M_w$. Similarly, if the concentration points at each angle are extrapolated to $c=0$ and these are then fit, the slope will be $<S^2>_z/3M_w$ and the intercept again $1/M_w$. Hence, the 'Zimm technique' allows determination of M_w, A_2 and $<S^2>_z$. The root mean square radius of gyration is often simply called the 'radius of gyration' and abbreviated as $R_g=<S^2>^{1/2}$. Nothing impedes the use of the Zimm approach in time dependent situations, as long as one is still operating within the range of validity of the approximations.

Figure 1 shows a Zimm plot from a batch experiment in which Automatic Continuous Mixing (ACM) was used to automatically vary the polymer concentration.[12] This technique involves much less manual work than the traditional method of discrete concentration measurements, and provides higher precision data. The continuously diluted sample flowed through both a multi-angle light scattering detector, a viscometer and an RI, where this latter detector allowed determination of c at each point.

[§] The so-called overlap concentration c^*, which marks the passage from dilute to semi-dilute solution is often approximated by $1/[\eta]$, where $[\eta]$ is the intrinsic polymer viscosity.

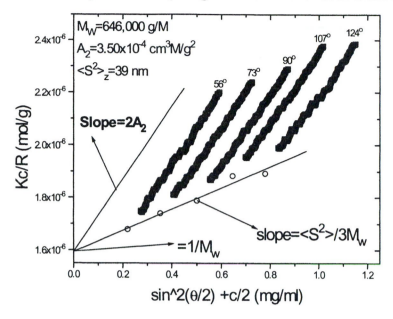

Zimm plot for PVP (using Automatic Continuous Dilution)

M_W=646,000 g/M

A_2=3.50×10⁻⁴ cm³M/g²

$<S^2>_z$=39 nm

Slope=2A₂

=1/M$_w$

slope=$<S^2>$/3M$_w$

Kc/R (mol/g)

sin^2(θ/2) +c/2 (mg/ml)

56° 73° 90° 107° 124°

Figure 1. Batch Zimm plot of poly(vinylpyrrolidone) using Automatic Continuous Mixing (ACM). Simultaneous viscosity measurements yielded [η] = 154cm³/g, and kₚ0.34. (Reprinted with permission from reference 12.) (Copyright 1999 Wiley.)

R(c,q) for the plot was computed from the sample scattering voltage V(c,q) according to

$$R(c,q) = \frac{V(c,q) - V_s(q)}{V_{ref}(q_r) - V_d(q_r)} N(q)R_{ref}F \qquad (12)$$

where $V_{ref}(q_r)$ and $V_d(q_r)$ are the scattering voltages of the calibration reference solvent and the dark voltage, respectively, at the reference angle $θ_r$, usually chosen as 90°. F in the above equation is an optical constant which accounts for two effects:** First of all, there is a reflection loss at each interface, and, secondly, because of refraction effects in two dimensions at the sample cell interface, the luminosity of the sample volume will appear smaller for solvents of higher index of refraction. Hence, if samples are measured in solvents different from that of the absolute calibration solvent, a correction for this effect must be made, which depends both on the cell geometry, and the solvents used. This effect has long been recognized.[13,14] While $R(c,q)$ could be determined at each value of q with respect to the reference solvent scattering at that q, practical considerations concerning the higher probability of stray light at angles away from 90^0 make use of a 'normalization' procedure preferable, where the normalization factor $N(q)$, is defined as

$$N(q) = \frac{V_n(q_r) - V_s(q_r)}{V_n(q) - V_s(q)} \qquad (13)$$

where $V_n(q_r)$ is the scattering voltage from the normalization solution at the scattering vector q_r that corresponds to the reference angle θ_r, $V_s(q_r)$ is the scattering voltage at q_r from the pure solvent the normalization solution is made in, and $V_n(q)$ and $V_s(q)$ are the normalization solution and pure solvent scattering voltages, respectively, at angle θ. A normalization solution contains Rayleigh scatterers, small, preferably massive particles, that scatter isotropically in the scattering plane. For aqueous measurements small latex spheres, e.g. of $R_g < 10nm$ are suitable, whereas in organic solvents polystyrene of $M < 50,000$ g/mole is suitable.

In practical terms, the two greatest historical banes of static light scattering have been 'stray light', and optical impurities in the polymer solution, such as 'dust'. Since the technique requires that the only light detected be that scattered by the polymers and solvent in the scattering volume, any stray light that comes from reflections and 'flare' from defects in the incident beam delivery optics (e.g. the beam entrance and exit windows, which may be scratched or

---

** There is another effect that is often ignored but that can be important when large scatterers are involved. Such particles scatter strongly at low angles, such that the very dim reflection of the laser beam off the exit window to air introduces an oppositely propagating beam with a supplementary angle form factor $P(180^0-\theta)$, which will yield increased scattering at high angles, with the appearance of a spurious scattering maximum at intermediate angles. The reflected beam amounts to roughly 4% for a typical air/glass interface, but is less for output windows specifically coated to minimize this effect.

have deposits of polymer or other impurities on them) can seriously distort the data. Since one normally subtracts the solvent scattering from the total solution scattering it is actually possible to cancel out stray light in the subtraction process, provided the flare light is steady and not too high. This type of subtraction, however, does not occur when using a pure solvent, such as toluene, as the absolute reference, so care must be taken to avoid any stray light in such procedures. A method of assessing stray light was recently introduced.[15]

The problem of 'dust' has been at least as problematic as stray light in the development of light scattering instruments and techniques. In the q=0 limit solid particles will scatter light in proportion to the sixth power of their diameter. This means, for example, that, assuming similar values of $\partial n / \partial c$, a 5 micron diameter dust particle will scatter on the order of 10^{18} times more light than a single 5 nanometer diameter protein molecule. Scattering voltages from 'dusty' solutions typically jump erratically and give meaningless results. Careful use of modern chemical filtration technology, however, can usually eliminate the dust problem. In SEC the columns themselves usually act as exceptionally high quality filters against spurious particles, so that light scattering data from SEC is typically very 'clean'.

Another strategy against dust is to make the scattering volume[††] very small so that individual impurity particles give large scattering 'spikes' in the data stream, which can then be digitally recognized and eliminated, or otherwise used. In fact, turning the impurity problem into a virtue, Heterogeneous Time Dependent Static Light Scattering (HTDSLS) was recently introduced with the explicit aim of being able to follow the number density of large scatterers, such as microbes, at the same time the properties of co-existing polymers, such as polysaccharides made or hydrolyzed by the bacteria, are characterized.[16]

It is important to point out that the above development is strictly applicable only to polymers for which $\partial n / \partial c$ is the same for all polymers in the population. Notoriously, copolymers produced by free radical polymerization have a range of comonomer compositions represented within the population. If the value of $\partial n / \partial c$ of the comonomers in polymeric form differ significantly from each other then meaningful measurements of the mass distribution cannot be made, unless scattering is measured in three different solvents in which $\partial n / \partial c$ of each component is different.[6,7] Use of the measured, average $\partial n / \partial c$ in a single solvent will yield on an apparent M_w. Similar problems exist with viscosity, because the viscosity of copolymers is not normally a simple weighted

---

[††] The 'scattering volume' refers to the portion of the illuminated sample volume that is actually detected, as opposed to the sample volume, which is the entire volume of solution in the scattering cell. Typically, scattering volumes in modern instruments are on the order of 10 nanoliters, whereas the sample volumes range from 10 microliters upwards.

average of the separate homopolymer values. For many water soluble copolymers the values of $\partial n / \partial c$ for each component often do not differ markedly, so that an approximately constant $\partial n / \partial c$ can be assumed.

Light Scattering in the SEC context

One motivation for moving the light scattering technique from batch to SEC is to obtain the entire population distribution associated with a polymeric population, as well as the associated characteristics such as dimensions and viscosity, instead of just the population averages, M_w and $<S^2>_z$ furnished by batch LS and viscometry. For example, this should allow the determination of the scaling law, if any, between M and $<S^2>$, a relationship of central importance in polymer science

$$< S^2 >^{1/2} \equiv R_g = BM^\beta \tag{14}$$

where $\beta=0.5$ for ideal random coils, ~0.6 for coils with excluded volume, 0.333 for spheres, etc. This type of scaling law is also of practical value in assessing polymer branching, especially when an unbranched polymer of the same type exists with which to make comparisons.[17,18,19]

In SEC practice dilute polymer solutions are separated and further diluted upon passing through the chromatographic column(s). This places the majority of SEC work in the dilute regime, where $1>>2A_2cM_w > >3A_3c^2M_w$, so that equation 11 again applies. The light scattering flow cells used in SEC pose no theoretical problems. The flow cell actually confers the special advantage of providing the exact same cell for all the solvent and sample measurements, thus eliminating the problems of the non-uniformity of multiple cells often used in discrete batch measurements. This is particularly important for eliminating small amounts of constant stray light and getting an accurate solvent scattering level under identical conditions as the polymer solution. On the other hand, the use of flow in dynamic light scattering can alter the nature of the autocorrelation function.[20]

Since SEC is performed in the dilute regime the $2A_2cM_w$ term can often be ignored, or a value of A_2 can be introduced in the analysis to make any small corrections needed. The dependence of A_2 (cm^3xMole/g^2) on mass is highly variable. In the good solvent limit A_2 is proportional to $M^{-0.2}$, whereas for branched structures it can be proportional to $M^{-0.5}$. If the dependence is known it can be introduced in the correction procedure. The error in the computation of mass, $\Delta M=M_{computed}-M_{correct}$, due to an error in A_2, the difference between the value used and the correct value, represented by $\Delta A_2=A_{2,used}-A_{2,correct}$ is

$$\frac{\Delta M}{M} = 2Mc\Delta A_2 \tag{15}$$

An underestimate of A_2 leads to an underestimate of M, and vice versa. Figure 2a shows 5% error contours of $\Delta M/M$ for M vs. c, for various values of ΔA_2. It is worthwhile to think of $\Delta A_2 = A_2$, i.e. figure 2a is the error caused in M when A_2 is ignored in the analysis. As seen, the effect can be quite large, so it is important to assess the conditions under which SEC is performed for any given polymer.

Because SEC provides the values of $c_i{}^{\ddagger\ddagger}$ and M_i, at every point i of the chromatogram, it is possible to construct any conceivable average, including the most commonly used ones; M_w, the number average mass M_n, and the z-average mass M_z. M_n is given by

$$M_n = \frac{\sum c_i \Delta v_i}{\sum \dfrac{c_i \Delta v_i}{M_i}} \tag{16}$$

where Δv_i is the elution volume increment between measurement point i and i+1, and is usually very small. For equal Δv_i, as is normally the case, Δv_i disappears from the above equation.

The denominator involves both c_i and M_i and is hence sensitive to the interdetector volume between the LS and RI detectors. The same applies for M_z, given by

$$M_z = \frac{\sum M_i{}^2 c_i \Delta v_i}{\sum M_i c_i \Delta v_i} \tag{17}$$

M_w, on the other hand, has the important property of being independent of the interdetector volumes, since

$$M_w = \frac{\sum M_i c_i \Delta v_i}{\sum c_i \Delta v_i} = \frac{\dfrac{1}{K}\sum R_i (q=0)\Delta v_i}{\sum c_i \Delta v_i} \tag{18}$$

and the numerator and denominator hence are equal to the entire sum of scattering and RI points, respectively, which are collected independently of each other. Here the $2A_2c$ term in equation 11 is assumed to be negligible. The same conclusion also applies if the $2A_2c$ term is kept and A_2 is constant, but no longer holds if A_2 is significantly mass dependent. An interesting feature of the term $\sum c_i \Delta v_i$ appearing in the denominator of equation 18 (and the numerator of

---

[‡‡] c_i represent the discrete concentration values yielded directly by periodic sampling from the concentration detector, which makes it amenable to finding M_n, etc. by summation. Some authors prefer to represent the concentration as a continuous function, such as $dC/d(\log M)$.

24

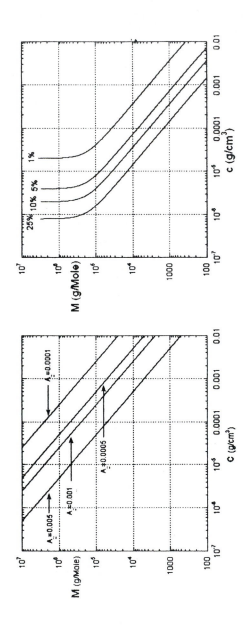

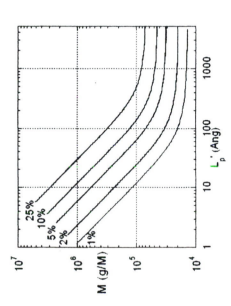

Figure 2a-c. Top left (a): Lines of 5% underestimate of M (equation 15), when A_2 is ignored (taken as zero) in the light scattering data analysis. Top right (b): Curves of constant root mean square error in M (<Err_M>, equation 22) corresponding to the value of M measured at concentration c, and the instrument characteristics defined in the text. Bottom (c): Error contours for persistence length/mass relationships when a single angle is used for LS. Here, $n=1.33$, $\theta=90^0$, $\lambda=677nm$, and $m/b=40$ g/MolexAngstrom.

equation 16) is that it is equal to the total mass injected, if 100% of the injected material passes through the column. Hence, if 100% of the material passes and the injected mass is accurately known, M_w can be computed with just a light scattering detector.

Another question that often arises is what minimum mass is measurable by light scattering. In fact, it is the combination of mass M, concentration c, and $\partial n / \partial c$ that determines detectability, not M alone. When working at the limits of detectability the $2A_2c$ term can generally be ignored so that $M=R/Kc$. In the following, the value of M at any elution point is computed corresponding to a concentration c, such that a root mean square fractional error in mass results $<Err_M>$, due purely to the limitations on precision of the LS and concentration detector (RI, UV/visible, etc.); i.e $<Err_M>$ is purely a function of the signal/noise ratio of the detectors, and systematic effects, such as slow drift in detector baselines, interdetector volumes, lateral broadening, and errors in $\partial n / \partial c$ that affect accuracy are ignored. $<Err_M>$ at each elution point is directly found by $M=R/Kc$ to be

$$< Err_M >= \frac{\delta M}{M} = \left[\left(\frac{\delta R}{R} \right)^2 + \left(\frac{\delta c}{c} \right)^2 \right]^{1/2} \qquad (19)$$

where the average of the covariant terms $<\delta R \delta c>$ is taken to be zero, since the random signal/noise fluctuations in the LS and concentration detectors are independent of each other. Here δR is the increment of R corresponding to the root mean square fluctuation of the LS baseline δV_{LS}. This can be approximated by computing the standard deviation of the scattering baseline of the LS instrument when pure solvent flows, a datum available at the beginning of any SEC chromatogram. Likewise, δc is the concentration increment corresponding to the RMS fluctuation of the concentration detector baseline δV_c:

$$\delta R = \frac{\delta V_{LS}}{S}, \quad \delta c = g \delta V_c \qquad (20)$$

where g is the calibration constant for the concentration detector, and S is the sensitivity of the detector, defined as Volts/Rayleigh ratio, which can be determined from the scattering voltage from the reference solvent (e.g. toluene) and the reference solvent's known Rayleigh ration R_{ref}.

$$S = \frac{V_{ref}}{R_{ref}} \qquad (21)$$

Typical values for a modern LS detector are $\delta V=0.001V$ and $S=2.25 \times 10^5$. Taking the examples of a typical water soluble polymer with $\partial n / \partial c =0.15$, an aqueous solution with $n=1.33$, and a diode laser with

$\lambda=677nm$, $K=1.25 \times 10^{-7}$. Figure 2b shows error contours for constant values of Err_M, computed according to

$$M = \frac{1}{Kc} \left[\frac{(\delta R)^2}{Err_M^2 - \left(\frac{\delta c}{c}\right)^2} \right]^{1/2} \quad (22)$$

that is, the x-axis gives the concentration needed to determine M on the y-axis to the percentage error shown on the corresponding error line. It is noted that c is the local detector concentration, not the injected concentration, which is much higher.

Another question that often arises is how many angles are needed to make a reliable scattering measurement. In principle, the more angles the better, since the effects of statistical errors associated with each one becomes smaller as the number of detected angles increases. If measurements are made on solutions containing particles such that $q^2 <S^2>_z <1$ then two angles would be the minimum requirement for extrapolation to $q=0$. If there is significant polydispersity, or if non-linear shape factors are involved, then two angles are insufficient, and even three or four angles are not likely to reliably illustrate a non-linear trend. A frequent case of severe non-linearity is when compact but massive aggregates co-exist with well dispersed single polymer chains.

For small particles a single angle measurement may suffice. Again, there is no hard limit to what size mass can be measured by a single angle. Rather, the error bars increase as $<S^2>$ of the particles increases. $<S^2>$ is related to mass via the shape and conformational properties of a polymer. For semi-flexible polymers with a definable persistence length L_p the 'wormlike' chain relation[21] is often used to relate the unperturbed $<S^2>$ to L_p and the contour length L of the molecule, which is itself usually directly proportional to M, via $L=M/(m/b)$, where m is the monomer mass and b the monomer contour length;

$$<S^2> = \frac{LL_p}{3} - L_p^2 + \frac{2L_p^3}{L} - 2(\frac{L_p^4}{L^2})\left[1 - \exp(-L/L_p)\right] \quad (23)$$

While $<S^2>$ in equation 23 refers to the dimension of the polymer in the absence of excluded volume interactions (often referred to as $<S^2>_0$), the perturbed value of $<S^2>$ is measured by light scattering, and most other techniques, so that it has become practical to speak of an 'apparent persistence length' L_p', which includes excluded volume effects, and use it in place of the unperturbed L_p in equation 23.[22]

Figure 2c shows selected error contours of M vs. L_p' for $\lambda=677nm$, $\theta=90^0$, n=1.33, and for m/b of 40 g/molexAngstrom, appropriate for sodium hyaluronate (HA), and selected errors; 1%, 2%, 5%, 10%, 25%. In the random coil limit, the relationship between M and L_p for a given fractional error, Err, becomes

$$M = \frac{3Err(m/b)}{q^2 L_p'}$$ (24)

which allows a quick estimate of errors for any values of m/b and q.

A final issue, related to the latter, is the minimum value of $<S^2>$ measurable. It is seen that when the $2A_2c$ term in equation 11 can be ignored, then

$$<S^2> = 3\frac{d[Kc/R(c,q)]/dq^2}{Kc/R(c,q=0)}$$ (25)

i.e. $<S^2>$, being proportional to the slope over intercept of $Kc/R(q,c)$ is independent of K (hence errors in $\partial n/\partial c$), of c (hence in errors due to the concentration detector), and even the calibration factor that goes into the determination of R. This means that the mean square error for $<S^2>$, $Err_{<S^2>}^2$ depends only on the fit to $Kc/R(c,q)$. This topic has been treated elsewhere,[23] so it is merely noted that a typical multi-angle LS instrument with $\lambda=677nm$ can measure R_g values with reasonable accuracy down to about 10 nm.

Fundamental Viscometry Notions

A fluid is said to be *viscous* when there is spatial inhomogeneity in the fluid's velocity field. Such inhomogeneity can arise both from differential momentum transport of fluid, and the presence of particles within the fluid that alter the velocity field, which contribute additional dissipative *internal friction* to the fluid. The amount of viscosity that a macromolecule contributes to a fluid is easy to measure, and can be related to conformational and other properties of the macromolecule, making viscosity measurements a valuable characterization tool.

Spatial variations in the velocity field are specified locally by the gradient term $\partial v_i/\partial x_j$, where v_i is the velocity component in the i-direction of a three component space and x_j is the jth directional component of the space. This gradient term has dimensions of s^{-1}, and is usually termed the *shear rate* $\dot{\gamma}$.

Stokes and Einstein[24] made the initial, arduous computations of the frictional factors of spherical particles in a fluid, and found the additional viscosity they contribute to be proportional to the fluid's own viscosity and a factor which is characteristic of the particle's own geometry and mass distribution, the *intrinsic viscosity* $[\eta]$. The total viscosity of a fluid, whose pure viscosity is η_s, and which contains particles of concentration c and having an intrinsic viscosity $[\eta]$ is given by

$$\eta = \eta_s \left\{ 1 + [\eta]c + k_p[\eta]^2 c^2 + O(c^3) \right\} \qquad (26)$$

where k_p is a constant related to the hydrodynamic interactions between polymer chains, usually around 0.4 for neutral, coil polymers,[25] and $O(c^3)$ represents terms of order c^3 and higher. It is customary to define the reduced viscosity η_r as

$$\eta_r \equiv \frac{\eta - \eta_s}{\eta_s c} = [\eta] + k_p[\eta]^2 c + O(c^2) \qquad (27)$$

$[\eta]$ is the extrapolation to zero concentration and zero shear rate of the reduced viscosity η_r. Determination of $[\eta]$ hence requires that η_s of the sample solvent and the total viscosity of the fluid containing the macromolecules η be measured. In SEC the concentration is usually low enough that it is assumed that $[\eta]=\eta_r$, where η_r is the quantity measured by combining viscometer and concentration detector data. Importantly, $[\eta]$ is a direct measure of the ratio of a polymer's hydrodynamic volume V_H to its mass M.

For example, $[\eta]$ for an ideal random coil in Θ-solvent conditions is[26]

$$[\eta] = \frac{\Phi_v}{M} \left(\sqrt{6} < S^2 >_0^{1/2} \right)^3 \qquad (28)$$

where $\Phi_v = 2.56 \times 10^{23}$. The value of Φ_v changes with chain architecture and the perturbing effects of hydrodynamic interactions.[21,27]

As for $<S^2>$, there is often a scaling law between $[\eta]$ and M, of the form

$$[\eta] = GM^\gamma \qquad (29)$$

where γ is sometimes termed the Mark-Houwink exponent. In view of equations 29 and 14, the R_g and viscosity exponents are expected to bear a relationship close to

$$\gamma = 3\beta - 1 \qquad (30)$$

Knowledge of G and γ also allows computation of the viscosity averaged mass M_η, which lies between M_n and M_w for $\gamma < 1$. Viscosity measurements can also be useful for assessing branching, since branched polymers will have smaller values of $[\eta]$ than their unbranched analogs.[17-19]

Viscosity Measurements

Several geometries can be used to create velocity gradients in fluids for the computation of η. The Navier-Stokes equation provides the basis for finding the relationship between η, the geometry and applied forces. One of the most common arrangements is the capillary viscometer, for which the Poiseuille solution to the Navier-Stokes equation is used:

$$\eta = \frac{\pi R^4 \Delta P}{8LQ} \tag{31}$$

where Q is the flow rate of solution through the capillary (in cm³/s) of radius R, across whose length L there is a pressure drop ΔP. If z is taken as the direction of fluid flow in the capillary and r is the distance from the center of the capillary then the shear rate is

$$\dot{\gamma}(r) = \frac{\partial v_z}{\partial r} = -\frac{4Qr}{\pi R^4} \tag{32}$$

The average shear rate in the capillary is found by integration over the capillary cross-section to be

$$\left| \dot{\gamma}_{ave} \right| = \frac{8Q}{3\pi R^3} \tag{33}$$

for which a typical value is 860 s⁻¹, given Q=1 mL/min, and R=0.0254cm.

Viscosity Measurements in the SEC Context

η_r can be computed directly from the voltage of a single capillary viscometer (a differential pressure transducer) at every point i, without need of an instrumental calibration factor, in terms of the viscometer baseline voltage V_b and the concentration at point i, c_i:

$$\eta_{r,i} = \frac{V_i - V_b}{c_i V_b}$$ (34)

This is so because the output of the viscometer is directly proportional to the pressure drop across the capillary of radius R and length L, which in turn is directly proportional to the total solution viscosity via Poisseuille's equation. The proportionality constant cancels out in the numerator and denominator of equation 34. Such single capillary viscometers are very inexpensive and simple to construct.

Alternatively, a Wheatstone bridge multi-capillary viscometer can be used to minimize fluctuations in pressure and other quantities. This leads to a significantly better signal to noise ratio than in the single capillary viscometer, but there are normally two separate differential pressure transducers, one of which measures a large pressure drop across the entire bridge, whereas the other measures a small pressure drop due to the small viscosity difference between the pure solvent viscosity in one capillary and the solvent with dilute macromolecules in another capillary. There is hence no cancellation of the calibration factors for the transducers, so that the calibration factor of each must be known. A thorough, simultaneous comparison of single capillary and bridge viscometers in SEC has been previously published.[28] A notable result was that, while the signal/noise of the bridge viscometer was always superior to that of the single capillary, the stochastic variations in the results between SEC runs exceeded the errors due to the signal/noise characteristics of the viscosity detectors.

Similarly to the LS case, where M_w is simply the ratio of the sums of the LS and concentration traces, the weight average intrinsic viscosity is equal to the ratio of the sum of the pressure differences over the peak, divided by the baseline value and the sum of the concentration.

$$[\eta]_w = \frac{\sum [\eta]_i c_i \Delta v_i}{\sum c_i \Delta v_i} = \frac{\sum \frac{V_i - V_b}{V_b} \Delta v_i}{\sum c_i \Delta v_i}$$ (35)

$[\eta]_w$ is hence independent of interdetector volume. Because the denominator $\sum c_i \Delta v_i$ is equal to the injected mass, if 100% of the injected material passes through the column, $[\eta]_w$ can be determined without a concentration detector if this condition applies, and the injected mass is accurately known.

Viscosity measurements have also been used to provide a 'universal calibration' for SEC columns, based on the notion by Benoit et al.[29,30] that if column separation actually occurs according to hydrodynamic volume, then viscosity/mass data can be combined to give V_H proportional to $M[\eta]$. With

M[η] plotted vs. elution volume, it is possible to compute the mass distribution of an unknown polymer with a viscometer and concentration detector, using the 'universal calibration' for the given column. This procedure is viable as long as no enthalpic, column adsorption, or any other effects are present that make the column separation mechanism deviate from separation according purely to V_H. In aqueous systems, for example, 'universal calibration/' often fails dramatically.[31,32,33] An example below illustrates the problem when separation does not occur according to V_H.

Errors in the determination of [η] depend on the signal to noise ratios of both the viscometer and concentration detectors. It is sometimes asserted that [η] can be determined to lower corresponding values of M than M itself can be determined by LS. This is based on the fact that LS measures Mc, whereas the viscometer measure $M^\gamma c$, where γ is usually smaller than 1, so that the viscometer signal drops off less sharply with decreasing M than LS. The actual errors, however, depend on signal/noise ratios, which are a function of the technical quality of each instrument, so that it is conceivable that for a set of LS and viscometer detectors, either one could outperform the other in terms of range of detectability and error level on M. Nonetheless, in terms of *dimensions*, [η] can easily measure equivalent hydrodynamic diameters down to a few tens of Angstroms, whereas, as mentioned, $<S^2>^{1/2}$ for LS typically has a lower limit around 100 Ang.

Interdetector Volume Effects

Unfortunately, even small interdetector volume uncertainties (on the order of 0.01mL) can cause dramatic uncertainties in the exponents β and γ, from above, as well as in other quantities, such as the polydispersity indices. This has been amply reviewed quantitatively in the past.[34] In addition to distortions in the analysis caused by interdetector volume errors, the interdetector volumes also produce 'lateral broadening' of the pulse of material as it proceeds through the detectors, for which corrective analyses exist.[35] The interdetector volume effects are large enough, in this author's opinion, that some of the original hope of precisely examining scaling laws and exponents, of the type in equations 14 and 29, for comparison to competing theories, has been tempered. Truly precision studies may require the advent of zero interdetector volume detectors; i.e. hybrid detectors that make all the relevant measurements in a single cell. While it is technically feasible to propose such a cell, the task of building a successfully operating one is a significant challenge. In the meantime, one means of assessing interdetector volumes is to remove the SEC columns, use a small injection loop (e.g. 10 microliters), inject any polymer desired, and set the detector sampling rate to at least 5 Hz. The volume difference between sharp detector peaks are more accurate than those made by injecting 'monodisperse' standards through columns, and attempting to match much broader peaks.

In spite of the nefarious effects of interdetector volume, both M_w and $[\eta]_w$ are independent of them, as shown above.

Example applications of multi-detector SEC

Analysis of a water soluble polymer, polyacrylamide (PAM)

Figure 3a shows the raw RI, LS and Viscosity traces for a polyacrylamide sample at 0.001 g/cm^3 in aqueous 0.1M NH$_4$NO$_3$ eluent, injected from a 0.1mL loop, using Shodex 804 and 806 columns in series, with a flow rate of 0.8 ml/min. The order of the detectors was viscometer to LS to RI. The raw data have been shifted by the interdetector volume differences between the detectors (0.10mL and 0.22mL, respectively). The fact that the peaks are in different positions is a reminder that the RI measures C, LS measures cM, and viscosity measures cM$^\gamma$. Figure 3b shows γ=0.69, so that the appearance of the viscosity peak between those of RI and LS is expected.

The inset to figure 3a shows M computed at each elution point and the concentration profile of the RI (the same as in the principal figure). These points are fit very well with a simple semi-logarithmic fit, which is the straight line in the figure. Although LS obviates the need for any column calibration, this type of fit is a 'hybrid' approach to data analysis. It extrapolates the absolute determination of M over the elution range where there is sufficient signal/noise in the LS data into the early and late ranges where the RI signal may still be good, but the LS data poor, as seen by the scattered M points at the beginning and end of the elution profile. This type of extrapolation is especially helpful at low values of M. This technique is self-consistent in that it uses a 'custom-calibration' curve for each experiment which applies identically to the polymer being characterized. This differs profoundly from the use of polymer standards of a certain chemical type and architecture to attempt characterizing other polymers of different chemical composition and architecture. It also is not dependent on a true size exclusion mechanism in the SEC process, as the 'universal calibration' described above is. As such, other than semi-logarithmic fits may often be required. For multi-modal populations this approach is not expected to hold across the whole elution spectrum, but fits over the sub-populations of similar architecture may be useful.

Figure 3b shows the analysis of the data from figure 3a plotted against M (as determined directly by light scattering, as opposed to including also the hybrid points, as described above). The concentration profile is uni-modal and quite broad, as the numbers in the graph reflect. The exponents β and γ are seen

34

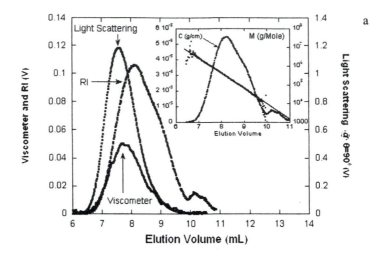

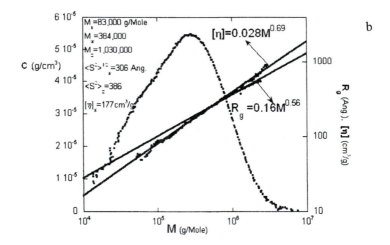

Figure 3a-b. Top (a): Raw GPC data for polyacrylamide showing RI, viscometer and LS at 90^0 The inset shows M both as determined directly from the LS and RI data, which includes the scattered points at each extreme of the elution spectrum, as well as the semi-logarithmic fit to these points, and the concentration data, for reference. (GPC data courtesy of Alina M. Alb). Bottom (b): Analysis of figure 3a data, including the concentration, R_g and [η] vs. M distributions. Also shown are the various average quantities from the analysis.

to closely obey equation 30. As pointed out, even small differences, e.g. 0.02mL, in the interdetector volumes can change these exponents significantly.

Origin of phase separation in gelatin/oligosaccharide solutions

It was found that mixtures of gelatin and oligosaccharides (acting as sweeteners) would sometimes spontaneously and unpredictably phase separate before they were processed into confectionery products. It was suspected that the gelatin was the culprit, since it is a complex, polyampholytic protein product whose detailed properties also depend on its source and method of preparation (e.g. acid vs. base extraction). Multidetector SEC of the oligosaccharides, however, provided the surprise answer to this puzzle: In some sources of oligosaccharide, produced by degradation of starches, there is a small population of long chain polysaccharides, which presents only a small shoulder on the RI SEC chromatogram but a very large light scattering peak where the shoulder lies, as seen in figure 4.[36] It was hence possible to determine the concentration and mass distribution of this tiny, long chain population in the oligosaccharides, and its presence led to considerable variations in M_w for oligosaccharides of identical M_n but from different manufacturers. Referring to the Flory-Huggins model for phase stability[37] it is found that, under athermal self-interaction in a polydisperse, multi-component system it is M_w of the polymers that controls the phase behavior, *no matter what the population distribution that leads to M_w.* Hence, a broad oligosaccharide population with a given M_w will give the same phase behavior as a narrow population to which a small amount of long chains are added to yield the same M_w.

Interestingly, the price on the market for the oligosaccharides was set by M_n, since the tongue's sense of sweetness is sensitive to the number of short chains it feels. In light of this work, however, the widely varying M_w is a critical concern for thermodynamic stability where oligosaccharides are used in multi-component systems.

Identification of Sub-populations and Properties of a Complex Polysaccharide

A good example of the power of multi-detector SEC is provided by 'gum arabic', a complex, multi-component natural product that exhibits large variations due to climactic, soil, and other conditions. Figure 5a shows a raw chromatogram of a typical sample of gum arabic in 0.1M NH_4NO_3 eluent and Shodex 804, 806 columns in series. A very rich, multi-modal profile is seen, with the contrast of the modes varying according to the detectors. The majority of the mass is seen in the RI peak at high elution volume, around 17mL, where

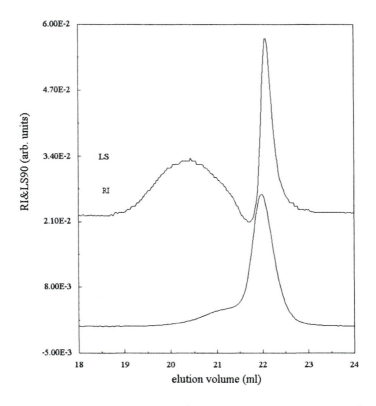

Figure 4. Raw RI and LS at 90° data for SEC of an oligosaccharide (dextran syrup). A small subpopulation of long chain polysaccharides is virtually undetectable by the RI signal (lower curve), but shows up strongly in the LS curve (upper). It is this small population, detected in this way, that leads to spontaneous phase separation in oligosaccharide/gelatin solutions. (Reprinted with permission from reference 36.) (Copyright 1997 Wiley.)

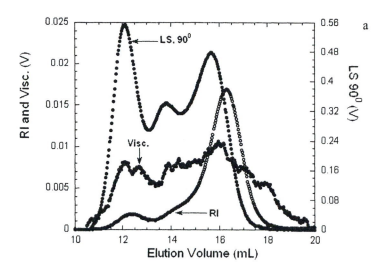

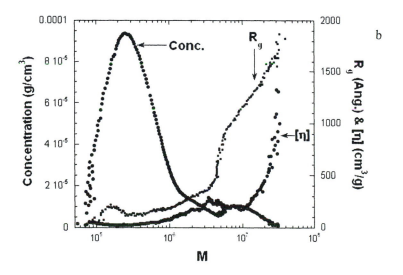

Figure 5a-b. Top (a): Raw LS at 90⁰, RI and viscosity data for SEC of gum arabic. The multi-modality is best seen in the light scattering signal. From A. Parker and W.F. Reed, unpublished results. Bottom (b): Analysis of the gum arabic data from figure 5a shows that most of the material is in a low mass, low viscosity form, whereas the small population of high mass polymer has qualitatively greater dimensions and [η].

there are also LS and viscosity peaks.[§§] The LS shows a middle peak for which neither the RI nor the viscometer yield peaks, whereas at low elution volume there is a very small RI peak, but highly pronounced LS and viscosity peaks.

Figure 5b shows the concentration, R_g and $[\eta]$ distributions vs. M. The bulk of the material is in the 'lower' mass, whose peak is around 250,000 g/mole, whereas there is a small peak of mass at high mass, whose peak is around 10^7 g/mole. This high mass fraction has exceptionally high R_g and $[\eta]$, and it is surmised that it is this tiny fraction that confers a marked viscosity to the material, and that the middle fraction, for which there is only a shoulder in the distribution, contains some protein (which shows up on chemical analysis, but was not detectable by the UV), which leads to a 'collapse' of the polysaccharide coil into a more compact structure, yielding lower R_g and $[\eta]$ than the high mass fraction. In fact a complete change in both R_g and $[\eta]$ behavior is seen to occur abruptly in going from the middle to high mass subpopulations. An SEC analysis with just a concentration detector would have been unable to discover the properties of these different modes.

Example Where Lower Mass Elutes Before Higher Mass

Figure 6 shows the raw LS and RI chromatograms for a complex biological polymer, a proteoglycan (PG), before (labeled '0 min') and after hydrolysis by NaOH (labeled '15 min'). The proteoglycan consists of a protein backbone of mass about 400 Kg/Mole and many sidechains of 1 Kg/mole - 20 Kg/mole glycosaminoglycans (or GAGs, highly charged polysaccharides).[38,39] The action of NaOH is to strip sidechains from the protein backbone.[40] Before hydrolysis the intact PG elutes at low volume, as seen in both the RI and LS (labeled 'MALLS') signals, which closely follow each other. After hydrolysis, a large fraction elutes at higher volumes, around 20mL, seen in the main RI peak, with an asymmetric, much lower amplitude LS peak accompanying it. This corresponds to the GAGs, which have been stripped from the backbone, and have much lower masses than the original, intact PG. At close to 25mL it is striking to note a very large scattering peak, only for the hydrolyzed sample (in the regime where the RI shows a large solvent mismatch). In fact, this latter peak is the protein backbone, which is much more massive than the side chains, but elutes *after* them, due to its complicated polyampholytic interaction with the column. This example illustrates how SEC columns do not necessarily separate according to M, especially in aqueous phase, and especially with complex

---

[§§] There was also an ultra-violet detector (UV) in the detector train, which ran from LS to RI to UV to viscometer, where the UV had a very large internal volume, leading to the longitudinally diffused signal seen in the viscometer signal.

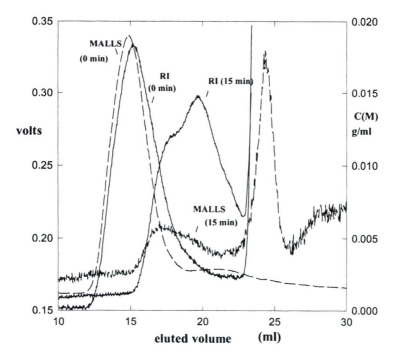

Figure 6. Raw RI and LS at 90° SEC chromatograms of proteoglycans before (0 min) and after hydrolysis by NaOH (15 min). Notice the large LS peak at 25mL for the backbone of the hydrolyzed PG. It is much more massive than the GAG chains that elute at the 20mL peak. (Reprinted with permission from reference 40.) (Copyright 1995 Wiley.)

mixtures of biological molecules of very different chemical nature. If deductions on a pure RI chromatogram were made using standard column calibration, they would be completely erroneous.

Alternatives to SEC

Although the power and versatility of SEC with multiple detectors has been well established over the years, often the slowness of the method, the complications of the columns and the instrumentation stability, together with the fact that SEC often provides more information than is needed for a given characterization task, makes alternatives attractive. It is hence important to take stock of recent developments in which the multi-detector system is kept, the SEC columns and injection loop are removed entirely, and different 'front-end' sample preparations are used. This has led to a family of techniques which is useful for both equilibrium and non-equilibrium characterization of polymer solutions. The basis for these techniques is Automatic Continuous Mixing (ACM). In ACM pumps and mixing devices are used to mix polymer solutions in any way desired; e.g. gradients of solute can be created to investigate properties of complex, multi-component systems in equilibrium or quasi-equilibrium, concentrated polymer solutions can be mixed with a solvent into a highly dilute regime, etc. Some examples are considered below.

ACM and equilibrium characterization

One of the simplest uses of ACM is to characterize two-component polymer solutions. This was first done for the case of PVP in aqueous solutions, in order to determine simultaneously M_w, A_2, A_3, $[\eta]_w$, and k_p (a portion of the results are seen in Figure 1).[12]

Another example is the effect that simple electrolytes have on semi-flexible polyelectrolytes. Figure 7 shows the raw LS and viscosity response of a linear polyelectrolyte, sodium hyaluronate (HA), to an increasing [NaCl] ramp, with [HA] fixed.[41,42] As expected, the light scattering increases since the electrostatically enhanced A_2, which suppresses scattering, decreases as ionic strength increasingly shields the charges on the HA, whereas the viscosity drops as the shielding leads to a contraction in the HA coil size. The inset to the figure shows A_2 and A_3 as determined from these and similar experiments. The often assumed proportionality of $A_3 \propto A_2^2$ is borne out by the data, which were also successfully analyzed in terms of electrostatic persistence length and excluded volume theories.[43,44,45,46]

A more complex application of ACM involves four component solutions containing solvent, polymers, surfactants, and simple salts.[47] Figure 8 shows how neutral, water soluble PVP behaves as a polyelectrolyte as the concentration of sodium dodecyl sulfate (SDS) increases at fixed [PVP] and no added salt. The decrease in light scattering and increase in viscosity demonstrate how the negatively charged SDS associates with the PVP causing its coil dimensions to increase, and hence also A_2 and $[\eta]$. After the PVP is saturated with SDS, any additional SDS simply increases the ionic strength of the solution, and leads to an increase in scattering and a decrease in viscosity. The inset to the figure shows how, at fixed [SDS] and [PVP] the A_2 of the polyelectrolytic complex decreases with increasing [NaCl], at the same time that r, the ratio of [SDS]/[PVP] increases as shielding allows a denser association of SDS to PVP.

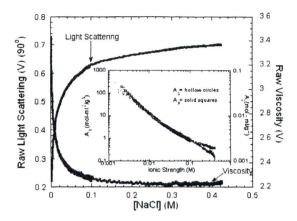

Figure 7. Raw LS data increases and viscosity data decreases for HA at 0.1mg/mL as it is ramped against a gradient of NaCl. The inset shows the values of A_2 and A_3 vs. ionic strength ([NaCl] in this), determined from these types of experiments. (Adapted from ref. 42.) (Copyright 2002 American Chemical Society.)

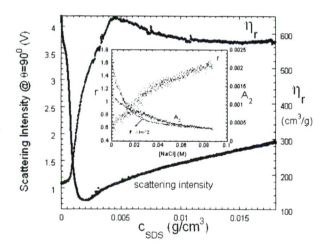

Figure 8. Raw data for a ramp of neutral PVP at 2 mg/mL vs. SDS, showing how LS is suppressed and viscosity increases as the SDS associates with PVP imparting it a negative charge and conferring polyelectrolyte properties on it. The inset gives the trends in A_2 and r (g of SDS associated/g of PVP) as NaCl is ramped at fixed concentration of PVP and SDS. (Adapted, with permission, from ref 47.) (Copyright 2002 American Chemical Society.)

Automatic Continuous Online Monitoring of Polymerization reactions (ACOMP).

The ability to follow polymerization reactions in real-time has many advantages, including i) understanding of fundamental mechanisms and kinetics involved in novel polymer syntheses, ii) optimizing reaction conditions at the bench or pilot plant level, and iii) providing integrated feedback control in full scale industrial reactors. This latter application is expected to bring significant improvements in product quality and considerable savings in non-renewable resources, energy, plant and personnel time.

Whereas several *in-situ* methods are currently used to obtain information on monomer and co-monomer conversion, such as near Infra-Red (NIR)[48,49,50] and Raman spectroscopy[51] most efforts at obtaining molecular mass information have centered on a literal adaptation of SEC systems to reactors. In these cases[52,53] the tendency has been to mimic the SEC operations that normally take place in the analytical laboratory; prepare discrete samples from reactor aliquots and inject these into an SEC system to obtain the usual information on mass distributions, and possibly conversion and viscosity. In an attempt to reduce the throughput time, shorter SEC columns have been used, or even eliminated entirely, in what is often termed 'flow injection analysis'.

ACOMP represents a conceptual break from the literal SEC and flow injection adaptations by providing a continuously diluted flow of sample to the multi-detector train, thus providing a continuous, detailed record of the reaction, and eliminating the complications, cycle of operations, expense, and dead times of the SEC and flow injection methods.

An example of ACOMP raw data is given in figure 9a.[54] This reaction involved the copolymerization of styrene and methylmethacrylate (MMA) in butyl acetate. The UV wavelength was set to 282nm, for which only styrene had significant absorption, and its decrease shows the disappearance of styrene as it is incorporated into copolymer. The RI signal is composed of the $\partial n / \partial c$ contributions from the styrene and MMA in both monomeric and polymeric forms. For both monomers $\partial n / \partial c$ is larger in the polymeric form, so the RI signal increases as conversion of monomer proceeds. In fact, the RI and UV data can be combined to obtain a continuous record of the composition distribution of the copolymer population, which is very difficult and tedious to obtain post-polymerization by other techniques, such as temperature rising elution fractionation (TREF),[55,56] and chromatographic cross-fractionation .[57,58] Such data taken for several experiments also allowed the determination of reactivity ratios. The continuous knowledge of the comonomer concentrations also allowed computation of M_w by LS, by being able to continuously integrate the composition equations involved in copolymer light scattering analysis. Figure 9b shows the type of composition profiles obtained by ACOMP for different

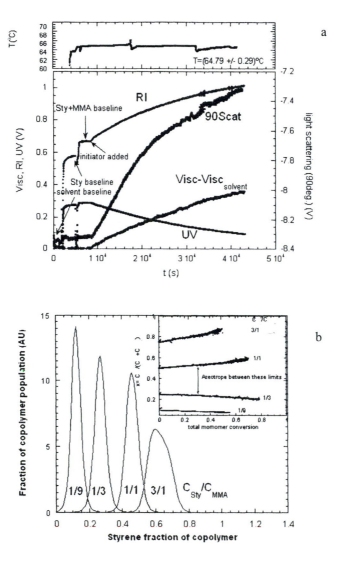

Figure 9a-b. Top (a): Raw ACOMP data; copolymerization of MMA and styrene. The behavior of each signal is described in the text. (Reprinted from reference 54 with permission) Bottom (b): Final copolymer composition distribution from data such as in Figure 9a, for reactions with different starting compositions of styrene/MMA (g/g). Inset shows the fraction of styrene remaining in the reaction at each moment of total monomer conversion. (Reprinted from reference 54 with permission.) (Copyright 2002 American Chemical Society.)

starting ratios of styrene/MMA. The inset shows the fraction of styrene in the remaining monomer at each instant.

Figure 10a is an illustration of the type of detailed monomer conversion kinetics obtainable by ACOMP,[59] in which the fractional monomer conversion of acrylamide (Aam) in aqueous solution is shown for varying amounts of persulfate initiator. At high amounts of initiator the conversion is nearly perfectly first order, following the predictions for free radical polymerization of the Quasi Steady State Approximation (QSSA) in the limit where initiator lifetime is longer than the time for complete conversion,[60] whereas noticeable deviations are seen as less initiator is used and competition for radicals occurs from impurities, such as O_2. Figure 10b shows the behavior of M_w vs. conversion f, for the Aam reaction with different amounts of initiator. As predicted by the QSSA, M_w decreases linearly with f once the QSSA conditions are established early in the conversion, and M_w decreases with increasing initiator. Also shown in the figure are M_w points found by SEC on aliquots withdrawn during the reaction, which are in good agreement with the continuous ACOMP points.

While ACOMP confers the advantage of speed, continuous data and simplicity with respect to the literal SEC approach to monitoring, it does not directly furnish molecular weight distributions, the forté of SEC. There are, however, several means of obtaining polydispersity indices, and even distributions, using ACOMP, with no chromatographic columns in place. These methods have been detailed,[61] and an example is shown in Figure 11 involving free radical polymerization of PVP in which a 'booster shot' of initiator was added at f=0.2. M_w furnished by ACOMP at each point (or by SEC when aliquots are withdrawn at discrete points) is the cumulative value for all the dead chains that have accumulated until that point. The corresponding instantaneous value of M_w, $M_{w,inst}$, can be computed according to the methods of reference 61, and $M_{w,inst}$ is shown. $M_{w,inst}$ clearly shows that a population of larger mass (around 80,000) was being produced steadily in the beginning of the reactions, but then, precipitously, small chains of 20,000 begin to be produced as soon as the initiator boost is added. Interestingly, M_w only shows a monotonically decreasing behavior, in which it is not obvious that a bimodal population lurks. The insets to the figure show histograms constructed from $M_{w,inst}$ before the initiator boost (upper left), for which only the high mass mode is seen, and at f=0.9, after the initiator boost (upper left) for which the bimodality and dominance of the small mass population are seen.

ACOMP is currently undergoing significant expansion into new types of reactions, such as controlled radical polymerization, both by nitroxide mediation,[62] including 'gradient copolymerization',[63] Atom Transfer Radical Polymerization, stepgrowth polymerization, and steady state production in continuous reactors.[64]

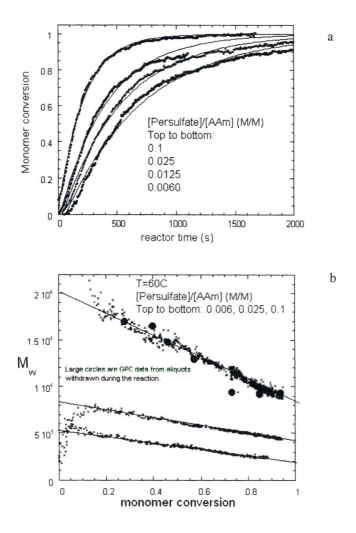

Figure 10a-b. Top (a): ACOMP conversion vs. time for PAAm, for different starting concentrations of persulfate initiator. The fits are to first order kinetics predicted by the QSSA for free radical polymerization. (Reprinted from reference 59 with permission). Bottom (b): M_w vs. conversion for the data from Figure 10a. The solid circles are SEC data points obtained on aliquots manually withdrawn during the reaction. The linear decrease in M_w after about 15% conversion conforms to the predictions of the QSSA in the limit where the initiator lifetime is long compared to total conversion time. (Reprinted from reference 59 with permission.) (Copyright 2001 American Chemical Society.)

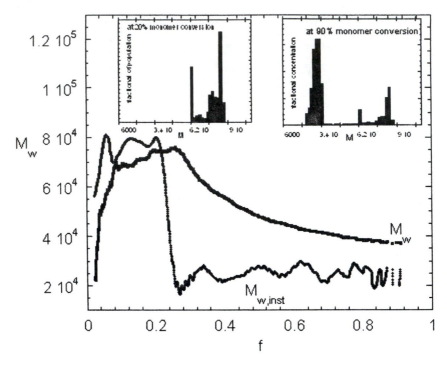

Figure 11. ACOMP results for M_w, and $M_{w,inst}$ vs monomer conversion f, for vinyl pyrrolidone polymerization, where an initiator 'booster shot' is added at f= 0.2. Insets show the large mass, monomodal population before the booster (upper left) and the bimodal population, with predominantly smaller mass (upper right) is shown at f=0.9. (From reference 61, with permission.) (Copyright 2000 American Chemical Society.)

Degradation, Dissolution, and Aggregation

The use of static light scattering to follow time dependent systems has been increasing to include degradation, aggregation, dissolution of dry and emulsified polymers, phase separation, etc.[65,66,67,68,69] This is sometimes termed Time Dependent Static Light Scattering (TDSLS). Often times these measurements are done simply in 'batch mode', whereas in other instances flow detectors are used.

Figure 12 shows an example of the dissolution of a polyelectrolyte, sodium Polystyrene sulfonate (PSS), in both pure water and water with 100 mM added NaCl.[70] The RI data in the inset show that there is no appreciable difference in dissolution rates whether the PSS is dissolved in pure water or salt

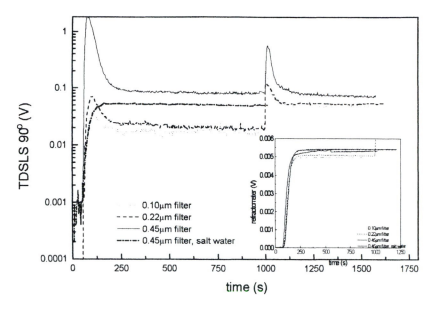

Figure 12. Dissolution of a polyelectrolite, NaPSS, in pure water and salt water with different types of filtration. In pure water a small population of large, dissolvable aggregates create a large scattering peak, whose height and subsequent plateau level depend on the filtration used. In salt water there is no scattering peak. The inset shows that there is no difference in terms of solubility rate in salt water and pure water. (From reference 70, with permission.) (Copyright 2000 Wiley.)

water (0.1M NaCl), and is also independent of the pore size of the inline filtration membrane. The TDSLS data, however, depend significantly on whether pure water or salt water is used, and also on the size of the filter. In pure water there are initial spikes in the light scattering, which are so large that a logarithmic scale is used to represent the TDSLS voltage. The larger the pore size, the larger the effect. The spike is due to a small population of transient aggregates present as the polyelectrolyte dissolves in pure water. The aggregates eventually disappear over a long enough period of time. These experiments strengthened earlier reports that the so-called 'slow mode' of diffusion[71,72] is due to removable aggregates,[73,74] rather than to any fundamental equilibrium nature of the polyelectrolyte solutions.

Simultaneous Multiple Sample Light Scattering (SMSLS)

There is currently a certain amount of emphasis on being able to rapidly screen new materials for applications in pharmaceuticals, structural materials, food products, etc. One approach in the SEC area has been to shorten the SEC columns, and obtain a higher serial sample throughput at the expense of chromatographic resolution. Other areas of high throughput screening involve instruments that can simultaneously measure many samples. Multi-channel plates have been adapted to UV/Visible spectrophotometry, infra-red photography, and several other techniques.

A recent development along these lines is SMSLS, where many samples can be monitored simultaneously by a single instrument.[75] This is an example of a non-ACM technique. SMSLS can be used to monitor the stability of many samples simultaneously against aggregation, phase separation, degradation, etc. over long periods of times (days, months), where it would be impractical to tie up a conventional light scattering instrument for a single sample. It can also be used to quickly monitor associations between particles, such as bioimmunoassays based on protein coated nanospheres interacting with antibodies. SMSLS is also useful for quickly ascertaining which reactions from a series of polymerization reactions meet certain criteria, such as whether polymerization occurs, and, if so, what the time scale is, roughly what the mass of the polymer is, and whether there are any particular features of the TDSLS signature that indicate certain mechanisms are occurring.

Conclusions

The principles and usual approximations behind the SEC practice of light scattering and viscosity have been presented, along with a number of examples that demonstrate the power of multidetector SEC with respect to conventional column calibration SEC.

In fact, SEC represents just one of several powerful configurations that can be used in conjunction with multidetector trains. While the role of multidetector SEC is well established and often provides a benchmark for new methods, it seems probable that in the near future the kindred methods of ACM, ACOMP, and SMSLS will come into widespread use, thereby widening the types of macromolecular phenomena susceptible to quantitative analyses.

Acknowledgments

Support by NSF CTS 0124006 and NASA NAG-1-02070 and NCC3-946 is gratefully acknowledged. The collaboration of many people over the years has been vital. These include Alan Parker, Jean-Luc Brousseau, Alina M. Alb, Gina A. Sorci, Michael F. Drenski, David P. Norwood, Emmanuel Mignard, Fabio Florenzano, Ricardo Michel, and others.

References

[1] Lord Rayleigh, *Phil. Mag.*, **1871**, *41*,107
[2] Lord Rayleigh, *Phil. Mag.*, **1899**, *47*, 375
[3] Einstein, A. *Ann. Phys.*, **1910**, *33*, 1275
[4] Debye, P. *J. Phys. and Coll. Chem.*, **1947**, *51*, 18
[5] Zimm, B.H. *J. Chem. Phys.*, **1948**, *16*, 1093
[6] Bushuk, W.; Benoit, H. *Can. J. Chem.*, **1958**, *36*, 1616
[7] Stockmayer, W.H.; Moore, L.D.; Fixman, M.; Epstein, B.N. *J. Polym. Sci.*, **1955**, *16*, 517
[8] Jackson, J.D. *Classical Electrodynamics*, 2nd Ed., John Wiley & Sons, NY, **1975**, ch. 9
[9] Brandrup, J.; Immergut, E.H., Eds. *Polymer Handbook*, 3rd Ed., **1989**, John Wiley & Sons
[10] Berry, G.C. *Adv. Polym. Sci.*, **1994**, *114*, 233
[11] Kerker, M. *The Scattering of Light and Other Electromagnetic Radiation*, **1963**, Academic Press, NY
[12] Strelitzki, R.; Reed, W.F. *J. App. Polym. Sci.*, **1999**, *73*, 2359
[13] Doty, P.M.; Zimm, B.H. ; Mark, H. *J. Chem. Phys.*, **1945**, *13*, 159
[14] Ahad, E.; Jennings, B.R. *J. Phys. D: Appl. Phys.*, **1970**, *3*, 1059
[15] Florenzano, F.H.; Strelitzki, R.; Reed, W.F. *Macromolecules,* **1998**, *31*, 7226
[16] Schimanowski, R.; Strelitzki, R.; Mullin, D.A.; Reed, W.F. *Macromolecules*, **1999**, *32*, 7055
[17] Burchard, W. *Adv. Polym. Sci.*, **1999**, 143, 113
[18] Zimm, B.H.; Stockmayer, W.H. **1949**, *17*, 1301
[19] Cotts, P.M.; Guan, Z.; McCord, E.; McClain, S. *Macromolecules*, **2000**, *33*. 6945
[20] Berne, B.J.; Pecora., R. *Dynamic Light Scattering*, **1990**, R.E. Krieger, Malabar, Fla.
[21] Yamakawa, H. *Modern Theory of Polymer Solutions*, **1971**, Harper and Row, N.Y., Sects. 9c, 14, 15
[22] Reed, C.E.; Reed, W.F., *J. Chem. Phys.*, **1991**, *94*, 8479
[23] Reed, W.F. ch. 2, pp. 7-34, in *Strategies in Size Exclusion Chromatography* **1996**, ACS ser. 635 , Eds. Potschka, M; Dubin, P.
[24] Landau, L.D.; Lifshitz, E.M. in *Fluid Mechanics*, **1959**, Pergamon Press, London

50

25 Huggins, M.L., *J. Am. Chem. Soc.,* **1942**, *64*, 2716
26 Flory, P.J.; Fox, T.G. *J. Am. Chem. Soc.*, **1951**, *73*, 1904
27 Kirkwood, J.G.; Riseman, J. *J. Chem. Phys.*, **1948**, *16*, 565
28 Norwood, D.P.; Reed, W.F. *Int. J. Polym. Ana. and Char.,* **1997**, *4*, 99
29 Benoit, H.; Grubisic, C.; Rempp, P.; Decker, D.; Zilliox, J.G. *J. Chim. Phys. Phys.-Chim. Biol.*, **1966**, *63*, 1506
30 Grubisic, C.; Rempp, P.; Benoit, H. *J. Polym. Sci., Polym. Lett. Ed.,* **1967**, *5*, 753
31 Dubin, P.L; Edwards, S.L.; Mehta, M.S. *J. Chromatogr.*, **1993**, *635*, 51
32 Bahary, W.S.; Jilani, M. *J. App. Polym. Sci.*, **1993**, *43*, 1531
33 Dubin, P.L.; Principi, J.M. *Macromolecules*, **1989**, *22*, 1891
34 Reed, W.F. *Macromole Chem & Phys*, **1995**, *196*, 1539
35 Wu, C.S. Ed. *Handbook of Size Exclusion Chromatography*, **1995**, Marcel Dekker, pp. 12,13,119-121
36 Vinches, C.; Parker, A.; Reed, W.F. *Biopolymers*, **1997**, 607
37 Flory, P.F. *Principles of Polymer Chemistry*, **1953**, Cornell University Press, Ithaca
38 Heinegard, D.; Paulsson, M. *Extracellular Matrix Biochemistry*, **1984**, Elsevier, NY
39 Ruoslahti, E. *Ann. Rev. Cell. Biol.*, **1988**, *57*, 375
40 Ghosh, S.; Reed, W.F. *Biopolymers*, **1995**, *35*, 435
41 Bayly, E.E.; Brousseau, J.-L.; Reed, W.F. *Int. J. Polym. Char. & Ana*, **2002**, *7*, 1
42 Sorci, G.A.; Reed, W.F. *Macromolecules*, **2002**, 35, 5218
43 Odijk, T. *J. Polym. Sci. Phys. Ed.*, **1977**, *15*, 477
44 Skolnick, J.; Fixman, M. *Macromolecules*, **1977**, *10*, 9444
45 Fixman, M.; Skolnick, J. *Mcromolecules*, **1978**, *11*, 8638
46 Reed, C.E.; Reed, W.F. *J. Chem Phys*, **1991**, *94*, 8479
47 Sorci, G.A.; Reed, W.F. *Langmuir*, **2002**, *18*, 353
48 Long, T.E.; Liu, H.Y.; Schell, B.A.; Teegarden, D.M.; Uerz, D.S. *Macromolecules*, **1993**, *26*, 6237
49 Aldridge, P.K.; Kelly, J.J.; Callis, J.B.; Burns, D.H. *Anal. Chem.* **1993**, *65*, 3581
50 Lousberg, H.H.A.; Boelens, H.F.M.; Le Comte, I.P.; Hoefsloot, H.C.J.; Smilde, A.K. *J. App. Polym. Sci.*, **2002**, *84*, 90
51 Ozpozan, T.; Schrader, B.; Keller, S. *Spectrochim. Acta A*, **1997**, *53*, 1
52 Ponnuswarmy, S.; Shah, S.L.; Kiparissides, C. *J. Liquid Chromatography*, **1986**, *9*, 2411
53 Budde, R.; Reichert, S. *Angew. Makromole. Chemie*, **1988**, *161*, 195
54 Giz, A.; Oncul Koc, A.; Giz, H.; Alb, A.M.; Reed, W.F. *Macromolecules*, **2002**, *35*, 6557

55 Feng, Y.; Hay, J.N. *Polymer*, **1998**, *39*, 6723

56 Wignall, G.D.; Alamo, R.G.; Ritchson, E.J.; Mandelkern, L.; Schwahn, D. *Macromolecules*, **2001**, *34*, 8160

57 Schunk, T.C.; Long, T.E. *J. Chromatog. A*, **1995**, *692*, 221

58 Yau, W.W.; Gillespie, D. *Polymer*, **2001**, *42*, 8947

59 Giz, A.; Catalgil-Giz, H.; Brousseau, J.-L.; Alb, A.M.; Reed, W.F. *Macromolecules*, **2001**, *34*, 5, 1180

60 Dotson, N.A.; Galvan, R.; Laurence, R.L.; Tirrel, M. *Polymerization Process Modeling*, **1996**, VCH NY

61 Reed, W.F. *Macromolecules*, **2000**, *33*, **7165**

62 Chauvin, F.; Alb, A.M.; Bertin, D.; Tordo, P.; Reed, W.F. *Macromol Chem Phys*, **2002**, *203*, 2029

63 Mignard, E.; LeBlanc, T.; Guerret, O.; Bertin, D.; Reed, W.F. *Macromolecules*, **2004**, *37*, 966.

64 Grassl, B.; Reed, W.F. *Macromol Chem Phys*, **2002**, *203*, 586

65 Holthoff, H.; Egelhaaf, S.U.; Borkovec, M.; Schurtenberger P.; Sticher, H. *Langmuir*, **1996**, *12*, 5541

66 Wright, L.S.; Chowdhury, A; Russo, P. *Rev. Sci. Instrum.*, **1996**, *67*, 10, 3645

67 Egelhaal, S.U.; Schurtenberger, P. *Rev. Sci. Instrum.*, **1996**, *67*, 2, 540

68 Norisuye, T.; Shibayama,M.; Nomura, S. *Polymer*, **1998**, *39*, 13, 2769

69 J.L. Ganter and W.F. Reed, *Biopolymers*, **2001**, *59*, 226

70 Michel, R.C; Reed, W.F. *Biopolymers*, **2000**, 53, 19

71 Lin, S.C.; Lee, W.I.; Schurr, J.M. *Biopolymers*, **1978**, *17*, 1041

72 Drifford, M.; Dalbiez, J.-P. *Biopolymers*, **1984**, *24*, 1501

73 Ghosh, S.; Peitzsch, R.M.; Reed, W.F. *Biopolymers*, **1992**, *32*, 1105

74 Smits, R.G.; Kuil, M.E.; Mandel, M. *Macromolecules*, **1994**, *27*, 5599

75 Drenski, M.F.; Reed, W.F. *Polymeric Mat: Sci & Eng*, **2003**, *88*, 304

Chapter 3

Characterization of Polymer Chain Architecture: Size-Exclusion Chromatography with Intensity and Dynamic Light Scattering and Viscometric Detectors

Patricia M. Cotts

Corporate Center for Analytical Sciences, DuPont Central Research and Development, Wilmington, DE 19880

The architecture of a polymer chain is critical to the properties of the polymeric material in film or bulk. The presence of branches of various types, backbone rigidity, polyelectrolyte effects or dense multimolecular aggregates all can be classified as architectural properties in solution that can dramatically affect bulk properties. Advances in on-line detection of light scattering (both static and dynamic) and viscosity enable characterization of polymeric architecture simultaneously with determination of molecular weight. Brief introduction and description of these light scattering and viscometric analyses are given, followed by specific applications. Particular emphasis is given to the most recent development of dynamic light scattering

Introduction

The vast majority of size exclusion chromatography (SEC, or equivalently, gel permeation chromatography, GPC) analyses are done to determine a polymer molecular weight and molecular weight distribution. Usually, a series of polymers of known molecular weight is analyzed and a curve relating the molecular weight to the elution time is constructed. The molecular weight distribution of polymers of interest are then evaluated from their elution time by reference to this curve. This works well when the polymers are the same, but series of polymers of known molecular weight are available for only a very few polymers. Even with these, concentration effects or column interactions can frequently affect elution times. For the many polymers not available as a molecular weight characterized series, the universal calibration is popular. An empirical correlation of elution time with a hydrodynamic molecular volume (defined as the product of intrinsic viscosity and molecular weight) has been shown to exist for a wide variety of polymers. Thus polymers of equivalent hydrodynamic volume are expected to elute at the same elution volume even if molecular weights are different. At the same molecular weight, more compact polymers (branched, or with heavier sidegroups) elute later than more extended polymers.

The intrinsic viscosity ($[\eta]$, expressed in dL/g or mL/g) is an indication of the spatial dimensions of the polymer molecule in dilute solution. The dependence of $[\eta]$ on the molecular weight M can be used to determine the dimensions of the polymer, and its architecture, especially when this information is available over a range of molecular weights. Alternative measures of the polymer size are the root-mean-square radius of gyration from the angular dependence of intensity light scattering, and the limiting diffusion coefficient obtained from dynamic light scattering. Direct determination of these parameters as fractionated polymers elute from an SEC column is a powerful tool for structural charaterization. A review of the theoretical relationships among these various size parameters, experimental data, and their dependence on polymer chain architecture is included in a recent book by Graessley.[1]

Frequently, determination of the intrinsic viscosity, $[\eta]$, is used to determine the molecular weight indirectly through use of a known Mark-Houwink relation, $[\eta]=KM^a$. If it is assumed that the universal calibration (the product $[\eta]M$ as a function of elution volume) is valid for a given polymer, and a viscosity detector is used to determine $[\eta]$, then M may be determined from the universal calibration curve, yielding M and $[\eta]$ independently across the distribution. This information may be used to assess architectural variations such as branching, stiffness or aggregate formation. Addition of a light scattering detector provides a direct determination of molecular weight that is not based on any empirical correlation with elution time. This presents a number of advantages, especially for architectural studies where subtle changes in size as a function of M are critical. Although Mark-Houwink relations for many polymers are reported in the literature, typically a large range of the parameters K and a is reported for a given polymer, and a linear relation is only valid over a limited range of M. Finally, use of the universal calibration can introduce systematic uncertainties that can be misleading for architectural determinations. Direct independent determination of both molecular weight and dimensions is greatly preferred.

Light Scattering

Intensity or Static Light Scattering

Classical light scattering has long been a primary tool for evaluation of polymer molecular weight. For a more detailed discussion, see the chapter by Wayne Reed in this volume. A recent book on polymer solutions by Teraoka also provides an excellent introduction to this topic.[2] Excess scattering (minus solvent scattering) is determined for polymer solutions as a function of scattering angle and concentration, and data is reduced using a Zimm plot or other means to extrapolate to zero scattering angle and concentration:

$$\frac{Kc}{R_\Theta} = \frac{1}{M_w P(\Theta)}\{1 + 2A_2 c + \cdots)$$ (1)

where

$$P^{-1}(q) = 1 + \frac{q^2 R_g^2}{3}$$ (2)

with $K = 4\pi^2 n^2 (dn/dc)^2/N_A\lambda^4$, where dn/dc is the differential refractive index increment and the scattering vector $q = (4\pi n/\lambda)sin(\Theta/2)$. Symbols have their usual meaning, Θ is the scattering angle, A_2 is the second virial coefficient, and R_g is the root-mean-square radius of gyration. Equation 2 for the single chain form factor or internal structure factor $P(q)$, yields R_g independent of shape for $qR_g<1$. It should be noted that neither the dn/dc nor the concentration need by known to determine R_g for dilute solutions, whereas these must be known accurately to determine M_w. For light scatttering coupled with SEC, the dn/dc is conveniently determined by integration of the refractive index peak using the known mass of the polymer injected. This can be advantageous when small amounts of residual solvents or other contaminants need to be separated from the polymer. For $qR_g>1$, the form factors $P(q)$ for different architetectures begin to deviate from each other, as shown in Figure 1. These differences are often used to determine particle shapes for scattering at much higher q, as for neutrons and X-rays.

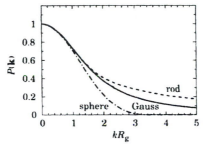

Figure 1. Particle form factor for spheres, random coils and rods. In this plot k is used to represent the scattering vector rather than q as in the text. From reference 2.

For light scattering, determination of R_g as a function of M is more useful as a determination of polymer chain architecture. While previously this required large amounts of polymer, tedious fractionation, and extensive analytical measurements, the availabilty of multi-angle light scattering instruments for coupling to an SEC has greatly advanced this capability. R_g can now frequently be determined over a decade in M on narrow fractions using less than 1 mg of polymer! Combining a few different molecular weights can extend this to several decades, which is important for semi-flexible polymers, as dicussed near the end of this chapter.

For flexible Gaussian polymers, the R_g determined by light scattering is a z-average, whereas the molecular weight determined by light scattering is a weight-average. This complicates determination of polymer dimensions for broad distribution polymers. This problem is greatly reduced when the polymers are separated using SEC. However, at a given elution volume, there is still a distribution of molecular weights, and the differing averages for R_g and M should be kept in mind in data interpretation. When used as a detector for SEC, light scattering is typically determined at very high dilution, often nearly an order of magnitude lower concentration than off-line light scattering. Thus, the second term in Equation 1 can usually be neglected, even in good solvents where A_2 is large. In off-line light scattering, excess scattering intensities are typically similar in magnitude to solvent scattering. With on-line light scattering, it is possible to determine excess scattering intensities that are only 5% of the solvent scattering. This is primarily due to the experimental advantages of the on-line instrmentation, in which the SEC column acts as a very efficient filter, the flow through cell eliminates extraneous scattering at the air-liquid interfaces, and the fixed photodiodes permit calibration of small variations in scattering volume.

When only a concentration detector, such as refractive index or UV-Vis, is used, the mass of the polymer injected onto the column is often chosen independent of the molecular weight. With light scattering and viscometric detectors, the response is nominally proprotional to the molecular weight. Thus, low molecular weight polymers must be analyzed at higher concentrations, and very high molecular weight polymers must be analyzed at very low concentrations. Fortunately, this is very consistent with the requirements of effiicient separation by size exclusion columns. In particular, very large polymers (very high molecular weight, or highly extended) must be injected at very high dilution to minimize overloading effects in separation. While these effects can be difficult to detect with only a concentration detector, with a light scattering detector, this is often detected as an inflection in the logarithmic dependence of the molecular weight as a function of elution volume. For very narrow distribution polymers, the width of the peak is characteristic of chromatographic band broadening rather than the molecular weight distribution, and the molecular weight from light scattering is usually independent of elution volume across the narrow peak.

Dynamic Light Scattering

Dynamic light scattering (DLS), also known as quasi-elastic light scattering (QELS) or photon-correlation spectroscopy (PCS) is the most recent addition to on-line detectors for SEC. For more detailed introduction to dynamic light scattering, a number of references are available.[2-5] The introduction of small solid state diode lasers, as well as avalanche photodiodes, and single-mode fiber optics, permits a tremendous reduction in size for instrumentation for dynamic light scattering. While typical off-line instruments reside on a large optical table, and use a motorized goniometer to access different scattering angles, the on-line instruments utilize fixed photodetectors, and are predominantly limited to a single scattering angle, typically 90 degrees.

Dynamic light scattering is widely used as a technique for particle sizing in the sub-micron range. These analyses are done on colloids and suspensions of particles which are compact and essentially rigid. Particle sizing instruments are often limited to 90 degreees scattering. As discussed below, while this often adequate for rigid particles, it is frequently not adequate for flexible chain polymers, where an angular dependence is often present, and sizes are small..

One of the advantages of dynamic light scattering is that it utilizes essentially the same instrumentation as intensity light scattering, requiring only the addition of a correlator and some modification to the detector optics. Close examination of the scattering intensity as a function of time reveals fluctuations that appear to be noise, but actually reflect the dynamics of the scattering entities. While intensity light scattering measures the average intensity, dynamic light scattering is used to determine the time scale of the fluctuations. In this sense, some requirements of the instrumentation are less stringent; the solvent scattering does not need to be subtracted, the scattering volume need not be known as accurately, and accurate concentration and dn/dc are not needed to obtain a particle size.

One modification to detector optics is the addition of an amplifier discriminator to the photomultipier tube (PMT) or avalanche photodiode (APD). This converts the analog scattering intensity into a series of photon pulses that can be counted by the correlator. The correlator then calculates the product of the scattering intensity at time t and $t+\tau$, as a function of τ.

$$\frac{\langle I(t)I(t+\tau)\rangle}{\langle I\rangle^2} = 1 + f_{coh}g_2(\tau)$$

For $\tau \to 0$, there is a high correlation in the intensity $I(t)$ and $I(t+\tau)$, and the initial value of the autocorrelation function is $\langle I^2 \rangle$. At long delay times, all correlation is lost, and the autocorrelation function decays to the square of the average scattering intensity, $\langle I \rangle^2$, as shown in Figure 2. The factor f_{coh} is the coherence factor, $0 < f_{coh} < 1$. A coherence area is a sufficiently small area on the photochathode surface so that the fluctuations being detected occur at a single point in the sample. If only one coherence area is viewed, all the light has the same phase, and $f_{coh} \to 1$. However, a single coherence area is very small, is difficult to achieve experimentally, and greatly reduces the scattering intensity. Conversely, viewing a large scattering area reduces f_{coh} to near 0. In practice, a compromise of a few coherences areas is typically used, and f_{coh} is generally 0.3-0.6.

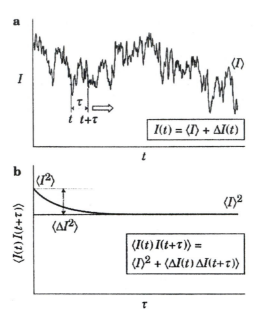

Figure 2. Part a: Fluctuations in scattering intensity; Part b: Autocorrelation function of fluctuations (from reference 2).

The normalized intensity autocorrelation function, $g_2(\tau)$, is then converted to the electric field autocorrelation function, $g_1(\tau)$:

$$g_2(\tau) = \left| g_1(\tau) \right|^2$$

For a dilute solution of monodisperse diffusing particles:

$$\left| g_1(\tau) \right| = \exp(-\Gamma\tau)$$

The particles exhibit random Brownian motion, and the electric field autocorrelation function decays as a single exponential, with a decay rate, Γ, given by:

$$\Gamma = Dq^2 \qquad\qquad (3)$$

and, for spheres:

$$R_H = \frac{k_B T}{6\pi\eta_s D}$$

where k_B is the Boltzman constant, T is the absolute temperature, and η_s is the solvent viscosity. The fluctuations in concentration reflect diffusion of *both* polymer (or particles) and diffusion of solvent. Thus D is a mutual diffusion coeffiicent, different from a tracer diffusion. In the limit of infinite dilution, both are equal to the center-of-mass diffusion coefficient of the particle, as long as there are no contributions of internal motions. The concentration dependence of the mutual diffusion coefficient reflects both the frictional properties and the thermodynamic polymer solvent interactions. Thus, in a good solvent, where the second virial coefficient is large, the mutual diffusion coefficent *increases* with concentration, in contrast to the tracer diffusion. At the very high dilutions used in SEC, this concentration dependence can typically be neglected.

The decay rate is linearly dependent on the square of the scattering vector, q, so that the timescale of the decay can be varied by changing the scattering angle. A smaller q can be advantageous for low molecular weight polymers, where very small sizes result in extremely fast decay rates when measured with standard 90 degree instrumentation. Equation 3 is strictly valid at high dilution, for monodisperse, spherical, rigid, non-interacting particles.

When used as an on-line detector with SEC, the restriction of high dilution is usually met. In fact, concentrations eluting from the SEC are typically much lower than those measured in off-line instruments. The separation of the distribution provided by the SEC columns produces a narrow distribution of sizes in each eluting fraction, so that the restriction of narrow polydispersity is reasonable. However, polymer molecules are typically neither spherical nor rigid. The long chain polymers are flexible, and internal flexibility gives rise to dynamics which also contribute to the dynamic light scattering. In this case, the center-of-mass translational diffusion, which is needed to calculate the hydrodynamic radius, is only obtained in the limit $q \to 0$. An extreme example of this is shown in Figure 3, for a polystyrene with molecular weight above 10 million. In this case, dynamic light scattering obtained at even lower scattering angles than 90°, such as 45°, yields a hydrodynamic radius, R_H, which is nearly a factor of two smaller than the limiting R_H at $q \to 0$.

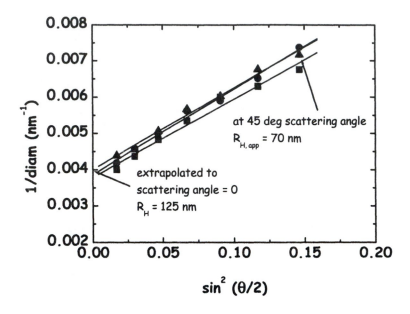

Figure 3. Dependence of 1/(2R$_H$) on scattering angle for a polystyrene with molecular weight of 13 million. Data from reference 6.

Dynamic light scattering is unique in that it provides a hydrodynamic parameter, the translational diffusion coefficient, from a non-hydrodynamic (i.e., a scattering) measurement. In many cases, hydrodynamic parameters such as the diffusion coefficient, intrinsic viscosity, etc., are measured directly. These parameters, often called transport properties, depend on movement of the particles, and are thus strongly affected by the movement of the solvent molecules in the vicinity of the polymer coil. This is referred to as hydrodynamic interaction, and varies with the degree of compactness of the polymer. The degree of hydrodynamic interaction is difficult to predict a priori, and thus parameters such as R_H or the intrinsic viscosity $[\eta]$, are more difficult to interpret than dimensions from direct scattering measurements such as R_g. However, these hydrodynamic parameters are often easier to determine accurately experimentally. Knowledge of more than one size parameter can also be a powerful means to assess the shape or conformation of the polymer, as discussed further below.

Polymer Solvent Interaction

Although most polymers are used in the bulk form, characterization of molecular parameters necessitates a dilute solution. In the bulk form, individual polymer molecules are intertwined with one another, and single molecules cannot be distinquished. An exception to this is isotopic labelling of a fraction of polymer chains in a melt of the polymer. In this case, neutron scattering can be used and interpreted as for a dilute solution in a solvent. In typical low molecular weight solvents, the polymer solvent interaction contributes to the spatial dimensions of the polymer. In an ideal or Θ-solvent, the polymer exhibits similar dimensions as it would dissolved in a melt of itself. These are referred to as unperturbed dimensions, and are appropriate for theoretical prediction of bulk properties, such as the Rouse model. For high molecular weight, linear, flexible polymers, these conditions are characterized by exponents of 0.5 in the dependences of a variety of size parameters, R_g, R_H, and $[\eta]$ on the molecular weight M. These ideal or θ-conditions are achieved when the second virial coefficient $A_2 \to 0$, and the polymer-solvent interaction is poor. These are called unperturbed dimensions; the polymer chain conformation is Gaussian, and can be adequately described by simple random walk statistics:

$$R_g^2 = \frac{N\ell^2}{6}$$

with the particle scatteirng factor is given by the Debye function:

$$P(q) = 2x^{-2}\left[1 - x^{-2}\left(1 - \exp\left(-x^2\right)\right)\right]$$

where $x=qR_g$.

The poor solvent conditions required to achieve θ-conditions are not easily compatible with column chromatography. Both the solvent and temperature must be carefully controlled, since in this region of the phase diagram, even small temperature changes can have a large effect on the polymer-solvemt interaction. In the limit of infinite molecular weight, the θ-temperature coincides with the critical solution temperature at which phase separation occurs. Thus these conditions are rarely used in SEC analyses, and measurements are usually carried out in thermodynamically good solvents for the polymers, where $A_2 >> 0$, and polymer-solvent interactions are favorable. An exception is so-called "critical chromatography", in which solvent conditons are carefully controlled to enable separation according to chemical composition rather than size. For SEC under more typical "good solvent conditions", individual polymer molecules are expanded or swollen to a larger size than would be obtained in the melt. These long-range interactions are more prevalent at higher molecular weights, so that the degree of expansion increases with M. For linear high molecular weight flexible polymers in good solvents:

$$R_g \approx M^{0.6}$$

and

$$[\eta] \approx M^{0.7}$$

These relations are valid for high molecular weight, linear, flexible polymers. As discussed further below, the exponents in the dependences of size parameters on molecular weight also reflect polymer chain architecture, and for hydrodynamic parameters, the hydrodynamic interaction. For these reasons, interpretation of limited data must be done with caution.

Applications

Branching in Polyolefins

One of the most useful and widely used applications of size exclusion chromatography to characterize polymer chain architecture is branching in polyolefins.[7] The long and short branches present in commercial polyolefins have tremendous impact on the degree of crystallinity and the rheological properties during processing and end-use. This rich area of research is beyond the scope of this chapter, but a few illustrative examples are given here. The short chain branches (generally hexyl and shorter) predominantly impact the degree of crystallinity. These can arise from intramolecular chain transfer in high pressure radical polymerizations (low density polyethylene, LDPE), or more recently, from co-polymerization with α-olefins such as hexene or octene in linear low-density polyethylene (LLDPE). These short branches have little impact on polymer backbone chain architecture, and are best characterized using spectroscopic techniques such as ^{13}C-NMR or IR. See the chapter in this volume by Paul DesLauriers for a more detailed discussion of combining SEC with IR for characterization of short-chain branching. The recent development of single-site metallocene catalysts has led to new levels of control in polyolefin branching. One intriguing example is the group of chain-walking catalysts developed by Brookhart and co-workers.[8] A group of polyethylenes synthesized using a palladium catalyst from this group is discussed below and compared with commercial semi-crystalline polymers.

The long chain branching, where branches can be as long as the main polymer chain, are more challenging to characterize. Even a small fraction of these branches can have a large impact on melt rheological properties, greatly influencing processing of these polymers. The presence of long chain branches is very difficult to detect by spectroscopic methods, and necessitates characterization of the polymer architecture.[1] A popular parameter to characterize degree of branching in polyolefins is g', defined as the ratio of intrinsic viscosity of a branched polymer to that of a linear polymer of the same molecular weight. This parameter as well as size ratios for R_g and others are frequently used to characterize branched architectures. Theoretical values for model branched polymers (stars, combs) are available in the literature.[1,9] The most popular model for the statistical branching in LDPE is that of Zimm and Stockmayer based on random non-linear polycondensation.[10] The determination of g' requires independent measurement of the molecular weight and intrinsic viscosity. These may be measured on the unfractionated polymer, or on fractions acquired by size exclusion chromatography or other means. In all cases, the polydispersity is a very significant issue. Polyolefins with long-chain branching are extremely broad in distribution, and comparison with a linear "standard" is problematic. Separation by SEC addresses some of these concerns, with $[\eta]$ and M determined continuously across the separated polymer. Frequently, this is done by reference to a "universal calibration" of the product $[\eta]M$ as a function of elution volume, as described above. It is preferable to acquire these measurements directly, using both a light scattering and viscosity detector. If the light scattering detector is a multi-angle detector, than the R_g may also be determined. The parameter g, defined as the ratio of the mean-square R_g for a branched polymer to a linear one of the same molecular weight, may then be compared with g'. Even in this case, polydispersity issues remain - at each "slice" of the chromatogram, there is still a distribution of molecular weight, and in the case of the branched polymer, a distribution of architectures as well. While R_g is not complicated by the varying hydrodynamic interaction that affect $[\eta]$, the experimental determination of R_g by light scattering yields a so-called z-average.

Thus, the R_g is highly dependent on the presence of even low degrees of polydispersity in the high molecular weight "slices", such as are prevalent in long-chain branched polyolefins.

Many of the studies of branching in polyolefins report a g' for polypropylene relative to polyethylene as an example of branching analyses. As discussed above, short-chain branches such as methyl groups are not amenable to this type of analysis, which assumes that the branches do not significantly perturb the backbone conformation. For polypropylene in particular, it is well-known that the addition of the methyl group favors the gauche conformation relative to polyethylene. This significantly shrinks the equilibrium dimensions of the polymer backbone. Longer or bulkier sidegroups present in high concentration may be expected to swell the equilibrium polymer dimensions due to steric effects. A detailed study of the effects of short-chain branching, using

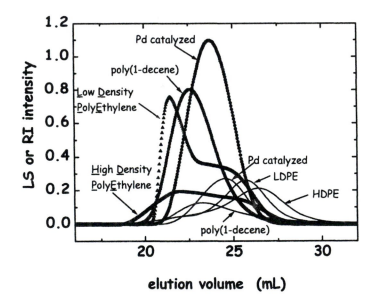

Figure 4. Refractive index (solid lines) and light scattering chromatograms (symbols) for a linear HDPE and 3 different types of branched polyethylenes, as indicated.

SEC with both light scattering and viscometric detectors, has been done by Sun and co-workers.[11]

Refractive index and light scattering (at 90° scattering angle) chromatograms for a variety of branched polyolefins are shown in Figure 4. The variety of branching architectures is already apparent from a qualitative examination of these chromatograms. The refractive index traces (solid lines) show that similar amounts of each polymer were injected. However, the branched polymers all exhibit a much higher light scattering intensity (dashed lines) than the linear polymer (high density polyethylene, HDPE), indicating much higher molecular weight at similar elution volume. The elution of the linear HDPE also begins much earlier, evidence of the much larger chain dimensions of the linear polymer, which will also be seen in the magnitudes of $[\eta]$ and R_g. Polymers with long-chain branching also have extremely broad polydispersity. Although the HDPE shown here is broad in molecular weight distribution, the LDPE is much broader. However, the LDPE elutes over a *narrower* range of elution volumes, due to the smaller hydrodynamic volumes of the branched polymers. Thus, even narrow "slices" of the LDPE are much more polydisperse in molecular weight and architecture than the the linear polymer to which they are compared. Finally, the shape of the light scattering chromatograms may be compared among the branched polymers. The LDPE, which has long branches predominantly at the high molecular weight end of the distribution, has a very asymmetric shape, while the poly(1-decene) with short octyl branches on alternate carbons and the Pd catalyzed polyethylene have a more symmetric shape, since branches are distributed more homogeneously throughout the molecular weight distribution.

Figure 5 shows the intrinsic viscosity as a function of molecular weight for "slices" across the distribution of the same polymers as shown in Figure 4. Using the HDPE as the linear standard, it is clear that the type of branching changes dependence of $[\eta]$ on molecular weight. The poly(1-decene), with short branches on every other carbon exhibits a linear dependence on M, as expected. The LDPE, with significant long chain branching, exhibits a strong curvature, as most of the long-chain branches are concentrated in the high molecular weight portion of the distribution. The Pd catalyzed amorphous polyethylene from the chain-walking catalyst, exhibits a linear dependence similar to the poly(1-decene). However, the short chain branching in this polymer includes substantial branch-on-branch structures. In this case, determination of the polymer chain architecture required both ^{13}C-NMR and SEC/LS/Viscometry, and was further verified using neutron scattering.[12]

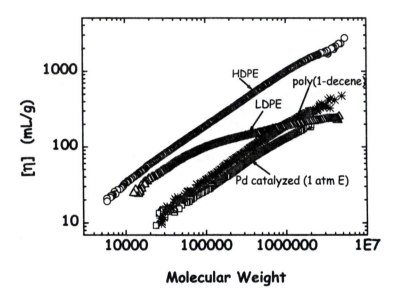

Figure 5. The intrinisc viscosity, [η], as a function of molecular weight for the same 4 polyolefins shown in Figure 4.

Star Polymers with Dynamic Light Scattering

Dynamic light scattering coupled with size exclusion chromatography is quite recent and applications are just beginning to appear, several of which are related to biological polymers such as proteins. As mentioned above, this scattering technique yields a hydrodynamic radius. For compact polymers, such as globular proteins, the size of the polymer in solution is quite small. Thus, a commonly used measure of size, the differential viscosity, may be difficult to detect at the high dilutions used in liquid chromatography. Also, the intrinsic viscosity is a measure of inverse particle density, and is insensitive to size changes for compact particles. Conversely, the hydrodynamic radius from dynamic light scatttering can be measured accurately down to a few nm. Compact materials scatter light highly relative to their size, improving the utility of dynamic light scattering for size measurement.

As discussed above, dynamic light scattering permits determination of a hydrodynamic radius simultaneously with the root-mean-square radius from intensity light scattering. An example of this for a broad distribution polystyrene is shown in Figure 6. Part A shows shows the angular dependence of the scattered light intensity near the maximum of the eluting peak, while Part B shows the autocorrelation function obtained simultaneously. This data was obtained using a Wyatt Technology Dawn EOS and the Wyatt QELS. It should be noted that at the "slice" indicated, the polymer concentration is about 0.03 mg/mL, or about 30 ppm, and for the dynamic light scattering, data is acquired for only 10 seconds per slice. For this measurment the QELS fiber optic was positioned at the photodiode #4, at 31.4° scattering angle. The measured R_g and R_H across the distribution are shown in Figure 7. The ratio R_g/R_H is about 1.65, as expected for linear flexible polymers. Table I below lists theoretical ratios of R_g/R_H for a variety of architectures. In general, more extended structures have larger R_g/R_H, while for the most compact structure, a hard sphere, $R_g/R_H <1$.

Table I. The ratio R_g/R_H for Various Polymer Architectures[9]

Architecture	R_g/R_H
Hard sphere	0.78
Random coil (Θ conditions)	1.50
Random coil (good solvent)	1.78
Star, f=4 (Θ conditions)	1.33
Star, f>>1 (Θ conditions)	1.08
Dendrimers (soft sphere)	0.98
Cyclic chain	1.25

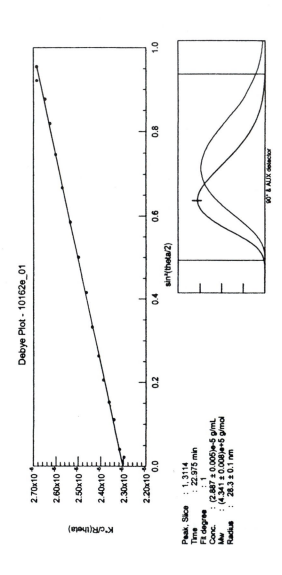

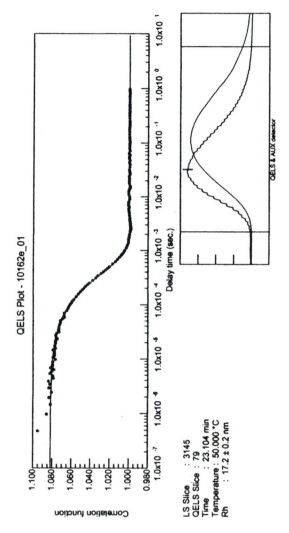

Figure 6. Part A (top): Intensity light scattering as a function of scattering angle at the "slice" of the distribution indicated. Part B (bottom): Dynamic light scattering correlation function obtained simultaneouly with QELS substituted for photodiode #4.

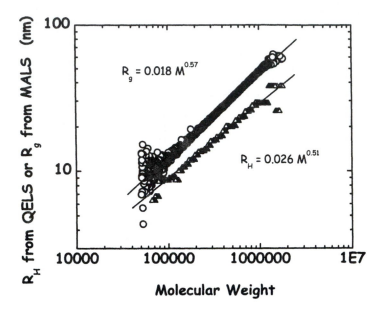

Figure 7. The R_g and R_H obtained for each slice across the distribution for the broad polystyrene shown in Figure 6.

As discussed above, branched polymers are more compact than linear polymers of the same molecular weight. The ratios of R_g and $[\eta]$ for branched and linear polymers of the same molecular weight were discussed above. These ratios require a linear standard of the same M. The ratios of R_g/R_H listed in Table I show that addition of two long branches (4 arm star) also decreases the R_g/R_H ratio in a Θ-solvent from 1.50 to 1.33. This ratio of two different radii can be obtained without reference to a linear standard. Two 3 arm stars of polystyrene were analyzed and results compared with those obtained for the linear polymer. These results, listed in Table II, were obtained in tetrahydrofuran (THF), a thermodynamically good solvent for the polymer, and are thus larger than the theoretical values in a Θ-solvent.

Even qualitative assessment of dimensions can be very valuable when characterizing polymers. For example, the presence of multi-molecular aggregates, micelles, incompletely dissolved polymer, or particulates can all occur in polymer solutions. All of these are more dense, and less solubilized than molecularly dispersed polymers. Typically, they scatter light strongly due to their high molecular weight, but often contribute minimally to the viscosity. Dynamic light scattering can be a powerful tool to determine the size of these species.

Table II. Comparison of R_g/R_H for Narrow Distribution Linear and 3-arm Polystyrenes

Architecture	M_w (MALS)	R_g (nm) (MALS)	R_H (nm) (QELS)	R_g/R_H	R_g/R_H (theor)
Linear	625,000	31.5	19.1	1.65	1.78
Linear	1,200,000	48.0	28.9	1.66	1.78
Star	250,000	20.5	13.8	1.48	1.33*
Star	650,000	33.4	22.7	1.47	1.33*

*for Θ conditions, f=4

Semi-rigid Polymers with Light Scattering and Viscometry

Highly extended polymers have larger R_g/R_H. An example is poly(hexylisocyanate) (PHIC), a well-studied polymer that adopts a helical conformation in dilute solution. Figure 8 shows a correlation function obtained for a collection time of 10 seconds, on a PHIC eluting from the SEC at a concentration of 0.44 mg/mL. Acquiring sufficient intensity for adequate statistics in the dynamic light scattering coupled with SEC is more difficult for this polymer for two reasons: 1) due to the highly extended configuration, the limiting concentration for efficient separation is lower than for a flexible polymer of similar molecular weight; 2) the *dn/dc* is quite small for this polymer in THF - improved data could be obtained in hexane, with a significantly larger *dn/dc*. Comparison of R_g and R_H across the molecular weight distribution is shown in Figure 9 for the PHIC. Although the R_H data are highly scattered, it is clear than R_g is much larger than R_H, as expected for a helical polymer.

For highly extended polymers, the intrinsic viscosity is often a more useful measure of the polymer dimensions in solution. Figure 10 shows the dependence of *[η]* on molecular weight for two samples of PHIC in THF. This data was obtained using SEC with both MALS and viscometric detectors. It can be seen that the data curves down at higher molecular weights. This is due to the semi-rigid nature of the helical PHIC in solution. At very low molecular weights, the polymer conformation would approache a rodlike limit, while at very high molecular weights, it approaches a random coil. Over limited ranges of molecular weight, the dependences appear linear, and can be fit using the familiar Mark-Houwink expression above. However, the exponent *a* will decrease with molecular weight. A more appropriate model is the wormlike chain. It should also be noted that this type of curvature appears similar to that

72

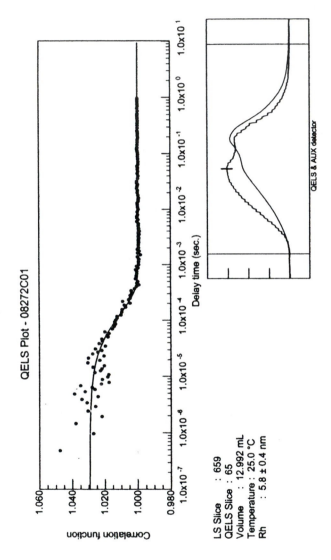

Figure 8. Correlation function obtained for PHIC in THF at the slice indicated using SEC/QELS.

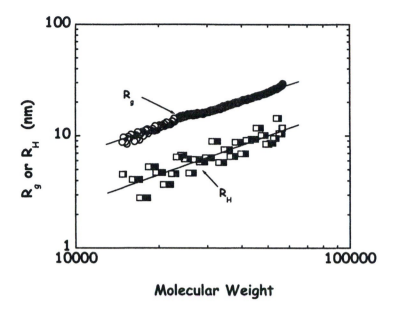

Figure 9. R_g and R_H as a function of M across the distribution of the PHIC shown in Figure 8 above.

observed when long-chain branching is present in polyolefins, as shown above in Figure 5. This is a clear example of the necessity of understanding the chemistry of the polymer before interpretation of the data.

Conclusions

The availability of multi-angle light scattering, dynamic light scattering, and differential viscosity detectors for SEC greatly increases the potential for rapid characterization of polymer chain architecture. The applications discussed here, long-chain branching, semi-rigid polymers, and the presence of multi-molecular aggregates are only a few of the possibilities. Applications are growing rapidly in the area of biopolymers and proteins, where these techniques are among few that can determine the amount of protein present in each oligomeric state. Studies of metabolic interactions, where the non-destructive determination of shape and size are critical, are also increasing. The combination of this architectural information with more commonly utilized spectroscopic techniques (UV-vis, IR) can be a very powerful tool in areas from biochemistry to luminescent polymers.

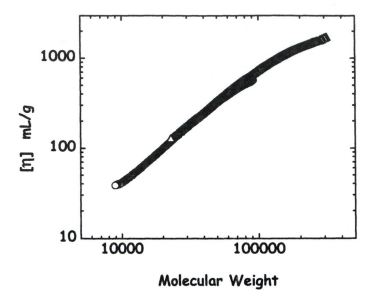

Figure 10. [η] as a function of molecular weight for fractions of PHIC eluting from the SEC. Two different polymers are shown. The curvature is due to the semi-rigid nature of the helical PHIC as discussed in the text.

References

1. Graessley W. W. 2004. *Polymeric Liquids and Networks: Structure and Properties,* London: Taylor and Francis.
2. Teraoka, I. 2002. *Polymer Solutions: An Introduction to Physical Properties,* New York: Wiley Interscience.
3. Berne B. J. and R. Pecora R. 1976. *Dynamic Light Scattering,* New York: Wiley Interscience.
4. Chu B. 1991. *Laser Light Scattering, Basic Principles and Practice,* 2nd Edition, San Diego: Academic Press.
5. Schmitz K. S. 1990. *An Introduction to Dynamic Light Scattering by Macromolecules,* San Diego: Academic Press.

6. Co C., P. M. Cotts, R. Burauer, R. de Vries, and E. Kaler. 2001. *Macromolecules* 34:3245.
7. Janzen J. and R. H. Colby. 1999. *J. Mol. Struct.* 485-486:569.
8. Ittel S. D. and L. K. Johnson. 2000. *Chem. Rev.* 100:1169.
9. Burchard W. 1999. *Adv. Polym. Sci.* 143:113.
10. Zimm B. and W. H. Stockmayer. 1949. *J. Chem. Phys.* 17:1301.
11. Sun T., P. Brant, R. R. Chance, and W. W. Graessley. 2001. *Macromolecules.* 34:6812.
12. Cotts P. M., Z. Guan, E. McCord and S. McLain. 2000. *Macromolecules.* 33:6945.

Chapter 4

Depolarized Multiangle Light Scattering Coupled with Size-Exclusion Chromatography: Principles and Select Applications

André M. Striegel

Department of Chemistry and Biochemistry, The Florida State University, Tallahassee, FL 32306–4390

The ability of polymers to depolarize incident light is important not only in end-use properties but also in the study of dilute polymer solutions. Quantitating the depolarization behavior of polymer solutions helps improve the accuracy of molar mass data derived from light scattering measurements, can provide information about changes in molecular rigidity as a function of molar mass, and can assist in the study of multi-component mixtures. To these effects, we have coupled depolarized multi-angle light scattering (D-MALS) on-line to size-exclusion chromatography (SEC). The principles of an SEC/D-MALS experiment are detailed, as are results of studies performed in our laboratory on various polymers, and of studies by other groups using batch-mode D-MALS.

The ability of polymers to depolarize light is important not only in end-use properties but also in the study of dilute polymer solutions. Polymers such as polystyrene, polycarbonate, and poly(vinyl butyral) are all extensively used in products governed by their optical performance characteristics. This is also the case for liquid crystalline polymers.[1] The case for studying polymers in dilute solution has been made quite eloquently by Burchard.[2] Quantitating the depolarization behavior of polymer solutions is important for a number of reasons: To improve the accuracy of molar mass data derived from light scattering measurements; to obtain information about the rigidity of polymers and the change in rigidity as a function of molar mass; to aid in the study of multi-component solutions in which the various components may have differing depolarization characteristics; etc.

To determine the molar mass-dependence of macromolecular rigidity and to separate the various components that may be present in a mixture, we benefit from the coupling of size-exclusion chromatography (SEC) to depolarized light scattering (D-LS). In this chapter we utilize the latter in multi-angle form, describe an SEC/D-MALS experiment and the information obtained from it, briefly contrast this technique to rheo-optical methods, and address both the advantages and shortcomings of SEC/D-MALS. The latter are done through the study of the semi-flexible polypeptide poly(γ-benzyl-L-glutamate) and the heavy-atom substituted polymer polystyrene bromide. We also show the successful application of SEC/D-MALS to the determination of impurities in a polymer, by taking advantage of the separation ability of SEC and of differences in depolarization behavior of sample and impurity, and describe batch-mode D-MALS experiments by other groups and some of the information derived therefrom.

Experimental

Materials

30 Kg/mol polystyrene (PS) standard was obtained from Pressure Chemical Co. (Pittsburgh, PA), poly(γ-benzyl-L-glutamate) [PBLG] from Polysciences (Warrington, PA), polystyrene bromide (PSBr) from American Polymer Standards (Mentor, OH), N,N'-dimethyl acetamide (DMAc) and LiCl were purchased from Fisher (Pittsburgh, PA). SEC/MALS experiments (performed in-house as described below) yielded the following molar mass averages for PBLG (all data in g/mol): $M_w = 148000 \pm 5000$, $M_z = 231000 \pm 10000$, $M_w/M_n = 1.96 \pm 0.09$; and for PSBr: $M_w = 674000 \pm 1000$, $M_z = 1170000 \pm 500$, $M_w/M_n = 2.07 \pm 0.03$. The differential molar mass distribution (MMD) of PBLG is shown as the solid line in Figure 1; that of PSBr as the solid line in Figure 2 (it should be noted that for PSBr this MMD was determined by SEC/D-MALS. See discussion below).

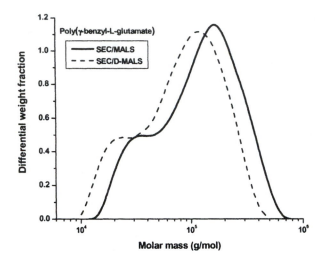

Figure 1. Differential molar mass distribution (MMD) of PBLG, as determined by SEC/MALS (solid line) and by SEC/D-MALS (dashed line). See text for explanation of shift.

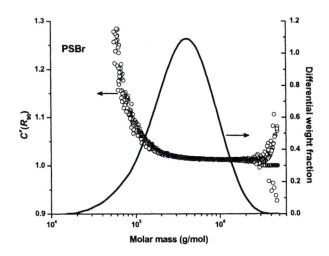

Figure 2. Distribution of the right-angle Cabannes factor, $C^v(R_{90})$, as a function of molar mass of PSBr (open circles), overlaid upon the polymer's MMD (solid line). Both parameters determined by SEC/D-MALS.

For preparing DMAc/0.5% LiCl, the salt was oven-dried overnight at 150 °C and maintained in a desiccator. After dissolving 5 g of LiCl in 1 L of DMAc at 100 °C, the solvent (DMAc/0.5% LiCl) was allowed to cool to less than 50 °C and then filtered through a 0.45 µm PTFE (Teflon) filter membrane (Phenomenex, Torrance, CA).

SEC/MALS

For SEC/MALS experiments, 25 mg of sample were dissolved in 10 mL of DMAc/LiCl by shaking in a laboratory shaker for 1 hour, then allowing to sit overnight, with gentle swirling. 400 µL of unfiltered solution were injected into a system consisting of a Waters 590 programmable HPLC pump (Waters, Milford, MA), a Shodex degassing unit (the mobile phase was also degassed by He-sparging in addition to vacuum degassing), a Waters 717+ autosampler, a DAWN EOS multi-angle light scattering photometer (Wyatt, Santa Barbara, CA), and an Optilab DSP interferometric differential refractive index detector (Wyatt). The detectors were connected in series with the refractometer last due to back-pressure considerations in this detector's cell. The detectors were maintained at 35.0 +/- 0.1 °C. Separation occurred over a column bank consisting of four analytical SEC columns (three PSS GRALlinear 10 µm columns and one PSS GRAL10000 10 µm column) preceded by a guard column (PSS Polymer Standards Service, Mainz, Germany). Column temperature was maintained at 35.0 +/- 0.1 °C with a Waters TCM column temperature system. Mobile phase was DMAc/0.5% LiCl at 1.0 mL/min. For all chromatographic determinations results are averages of four runs from two separate dissolutions, with two injections per dissolution.

The MALS detector, which measures scattered light at 17 different angles simultaneously, was calibrated by the manufacturer using toluene. Normalization of the photodiodes was performed in-house using a small, monodisperse ($M_w/M_n \leq 1.06$), isotropic scatterer, linear polystyrene with $M_p = 30$ Kg/mol. This PS was also used to determine the interdetector delays for SEC/MALS. Data acquisition and manipulation was performed using Wyatt's ASTRA for Windows software (V. 4.73.04).

The specific refractive index increment, $\partial n/\partial c$, of PBLG was determined to be 0.104 (± 0.001) mL/g, that of PSBr as 0.110 (± 0.001). This was done using the Optilab DSP refractometer off-line, under the same solvent/temperature/wavelength ($\lambda = 685$ nm) conditions as the SEC/MALS experiments, using Wyatt's DNDC for Windows software (V 5.20 (build 28)). In each case data were obtained from five dissolutions, ranging from 0.3-3.0 mg/mL.

SEC/D-MALS

In SEC using depolarized multi-angle light scattering (SEC/D-MALS), two strips of Polaroid film (Wyatt) are placed around the sides of the flow cell of the MALS unit and maintained tightly in place by a grooved cell retainer attached to the cell. One strip is vertically polarized and the other horizontally polarized. Data were initially taken with the former facing the odd-numbered photodiodes, the latter the even-numbered photodiodes. This procedure was then reversed. Normalization was effected in both modes with the 30 Kg/mol PS standard. The same dissolutions of sample and standard were analyzed in both polarization modes. Chromatographic conditions were the same as those detailed above. Minor flow rate fluctuations were corrected manually by using the retention times of the solvent/air peaks common to the refractometer traces of the samples.

Fundamentals and Caveats

Theory of SEC/D-MALS[3]

The incident light from the GaAs laser ($\lambda = 685$ nm) in the MALS unit is vertically polarized. As mentioned in the previous section, in depolarized multi-angle light scattering (performed as an on-line experiment with SEC, or SEC/D-MALS), the flow cell is covered with two polarizing filters (Polaroid strips). We first placed the filter with vertical transmission axis facing the odd-numbered photodiodes of the MALS unit and the filter with horizontal transmission axis facing the even-numbered photodiodes. After obtaining data in this mode, the placing of the filters was reversed. (It should be noted that the 90° photodiode is an odd-numbered photodiode). The filters play the role of analyzer in the optical train of the system. If the scattering particle is completely isotropic, the induced dipole will be parallel to the electric vector (E_0) of the incident light.[4] Assuming no absorption by the polymer, only the component of the scattered light parallel to the transmission axis of the analyzer (polarizing filters), $E_0\cos\phi$, will be passed on to the photodiodes. (Here, ϕ is the angle between the transmission and polarization axes, *not* the angle at which the photodiodes of the MALS detector are placed around the flow cell with respect to the direction of propagation of the laser light. The latter is given the symbol θ in the present chapter). The intensity of scattered light reaching the photodiodes is given by equation (1):

$$I(\phi) = \frac{c\varepsilon_0}{2} E_0^{\,2} \cos^2 \phi \tag{1}$$

where c is the speed of light in vacuum and ε_0 is the permittivity of free space. The intensity of the scattered light transmitted by the analyzer will reach a maximum, $I_{max} = \left(c\varepsilon_0 E_0^{\,2}\right)/2$, at $\phi = 0°$ and is zero when $\phi = 90°$. Equation (1) can be rewritten as equation (2):

$$I(\phi) = I_{max} \cos^2 \phi \tag{2}$$

in which form it is generally known as Malus's law.[5] Therefore, when an isotropic molecule in the flow cell scatters the (vertically polarized) light from the laser, none of the scattered radiation should reach the photodiodes facing the horizontal-polarization strip and all of the scattered radiation should reach the photodiodes facing the vertical-polarization strip. Another way of putting this is that the depolarization ratio at $90°$ should be zero. The ratio is given the symbol ρ_θ^v and is defined as:[6-9]

$$\rho_\theta^v = \left(\frac{I^v{}_h}{I^v{}_v}\right)_\theta \tag{3}$$

The superscript v in this paper will refer to the fact that the incident radiation is vertically polarized while the subscript θ refers to the scattering angle. $I^v{}_h$ and $I^v{}_v$ are the intensities of the horizontally- and vertically-polarized components of the scattered light, respectively. The above derivations relied on the assumption that the scattering particle was completely isotropic. If it is not, fluctuations in the orientation of the scattering particle offer the possibility of additional scattering, as the induced moment is generally not parallel to the electric vector of the incident light.[4] Factors that can contribute to improper measurement of the depolarization ratio (most of which do not have bearing on the present experiments, but are mentioned for the sake of completeness) include imperfections in polarizers and analyzers, fluorescence, optical activity, photodetector sensitivity to light polarization, multiple and stray light scattering, finite acceptance solid angle, and strains in sample cell windows.[9,18]

Cabannes showed that the excess scattering due to anisotropy is related to the depolarization ratio.[7] Thus, the Rayleigh ratio for a system of anisotropic particles, $R_\theta^{v,tot}$, is:

$$R_\theta^{v,tot} = R_\theta^{v,iso} \left(\frac{3+3\rho_\theta^v}{3-(4+7\cos^2\theta)\rho_\theta^v} \right) \tag{4}$$

where $R_\theta^{v,iso}$ is the Rayleigh ratio for isotropic particles. For a given observation angle θ.[8]

$$R_\theta^{v,tot} = R_\theta^{v,iso} C^v(R_\theta), \quad C^v(R_\theta) = \left(\frac{3+3\rho_\theta^v}{3-(4+7\cos^2\theta)\rho_\theta^v} \right) \tag{5a}$$

which reduces to (5b) at $\theta = 90°$:

$$R_{90}^{v,tot} = R_{90}^{v,iso} C^v(R_{90}), \quad C^v(R_{90}) = \left(\frac{3+3\rho_{90}^v}{3-4\rho_{90}^v} \right) \tag{5b}$$

The correction term $C^v(R_{90})$ is usually known as the Cabannes factor though in this chapter "Cabannes factor" denotes the more general $C^v(R_\theta)$. Both these terms will assume different forms when/if unpolarized incident radiation is used (Cabannes's original *facteur d'anisotropie*, calculated for the right-angle scattering of unpolarized incident radiation, equals $(6 + 6\rho^u)/(6 - 7\rho^u)$, where the superscript denotes the polarization state of the incident light).[6b,7,9] As the determination of molar mass *via* light scattering is dependent upon determination of the excess Rayleigh ratio of a polymer in solution, an increased accuracy in R_θ will lead to increased accuracy in the calculated molar masses.[4,10] Alternative explanations of depolarized light scattering, emphasizing different aspects of this technique, have been given in various references, *e.g.*, that by Jinbo *et al.*[11] For a more extensive review of anisotropic scattering and of the cooperative effects encountered in depolarized light scattering, the reader is referred to reference 12. The effects of anisotropy on light scattering measurements and resultant calculations have been dealt with in detail in reference 8b. For a comprehensive treatment of the polarization of light, see reference 5.

At this point we note that, while SEC/D-MALS may display a superficial resemblance to certain rheo-optical methods of analysis (most notably to flow birefringence), the type of information derived from the chromatographic and rheological methods is, at present, quite different. A discussion of techniques of the latter type is beyond the scope of this chapter; for comprehensive, in-depth review and discussion of rheo-optical methods, the reader is referred to the chapter by Lodge in reference 13, as well as to the monograph by Fuller,[14] among others.[15,16] The use of non-matching solvents (i.e., of solvents with a refractive index different from that of the polymer, or of polymer solutions with $\partial n/\partial c \neq 0$) in SEC/D-MALS, for obvious purposes, leads to a *form* birefringence caused by induced internal electrical fields of macromolecules, in addition to the *intrinsic* birefringence caused by preferential alignment of anisotropic materials in one direction.[16] The extreme difficulty in SEC/D-MALS of determining birefringence in the absence of form effects, as well as the equally difficult task of measuring an extinction angle,[13,14,16] precludes this technique from providing the type of information obtained from a flow birefringence experiment (*e.g.*, shear stress, first normal stress difference, etc.). Our aim is thus one of complementarity rather than replacement, and this will be expounded upon more in-depth in the Results & Discussion section.

Analyzer absorption and data treatment[18]

It has been noted that absorption of light by the Polaroid filters that act as analyzers in the SEC/D-MALS set-up results in an approximately 10-20% reduction in the intensity of the scattered light reaching the photodiodes of the MALS unit.[3] As a result, the MMD determined by SEC/D-MALS is identical in shape to that determined by SEC/MALS, though the former experiences a negative molar mass shift with respect to the latter. The results of this, for PBLG, may be seen in Figure 1. We proceed to outline a possible route for correcting the data from an SEC/light scattering experiment for this and other effects:

1. Data should first be corrected for minor flow rate fluctuations. While said fluctuations have a relatively minor effect in SEC/MALS, as compared to other techniques such as universal calibration, SEC[3], or relative calibrations, they are important in the present context to ensure that the same elution slices are being compared for depolarization purposes.

2. Absorption effects are determined based on (7):

$$(I^{v}{}_{Pol})_{\theta,i} = (I^{v}{}_{Tot})_{\theta,i} - (I^{v}{}_{h})_{\theta,i} - (I^{v}{}_{v})_{\theta,i} \tag{7}$$

where $I^{v}{}_{Tot}$ corresponds to the total intensity of scattered light (measured in SEC/MALS, i.e., in the absence of filters), $I^{v}{}_{Pol}$ is the amount of light absorbed by the Polaroid filters, and $I^{v}{}_{h}$ and $I^{v}{}_{v}$ are defined same as above. The subscripts i and θ indicate that the data are to be determined at each data slice i of each scattering angle θ being studied.

3. The molar mass of each elution slice is assigned based on the SEC/MALS data.

4. An analyzer correction factor is calculated for each elution slice (i.e., for each molar mass slice) at each angle, using the information from points 2 and 3.

5. The $C^v(R_\theta)$, used to correct the Rayleigh ratio for deviations due to anisotropy, are calculated from the SEC/D-MALS data at each molar mass slice at each angle.

6. The $C^v(R_\theta)$ are corrected for analyzer absorption effects (i.e., for molar mass shift) using the correction factors calculated in point 4.

7. Resultant data are incorporated into the proper MALS algorithm to yield properly flow rate-corrected, depolarization-corrected, analyzer absorption-corrected molar mass averages and distributions.

Results and Discussion

SEC/D-MALS of PSBr and PBLG

We now return to the question formulated earlier, namely of what use is SEC/D-MALS? The importance of determining the Cabannes factor for the purposes of accurate molar mass calculations from light scattering data was recognized at least as early as 1944 by Debye,[19] and early on Stacey pointed out that the causal relationship between anisotropy and depolarization provides a means of determining the form of the molecule.[20a] (Stacey also noted the depolarization behavior of particles larger than the wavelength of the incident radiation. Whether these particles are isotropic or anisotropic, they will show depolarization due to higher terms in the Mie equations). Stacey's observation was based on extensive earlier work, most notably by Krishnan, by Perrin, and by Zimm, Stein, and Doty, among others.[20] To quote Zimm et al., "...since the explanation of depolarization must lie in the fact that the scattering centers, the molecules, are asymmetric and anisotropic, it is reasonable to expect that, if the relation between cause and effect can be determined, depolarization measurements will contribute to our knowledge of the size and shape of molecules."[20b] In 1948 Doty plotted the approximate relation between the molar mass of a generic flexible polymer molecule and its depolarization factor, ρ^v,[21] and in 1953 Horn and Benoit used the anisotropic rod model to calculate the length of tobacco mosaic virus.[22] Much of the early work was also reviewed by Norris.[20e]

More recently our group noted that, for a particular set of solvent/temperature conditions, the depolarization behavior of dilute polymer solutions (measured as a function of molar mass) could differ due to either tacticity, heavy atom substitution, or aggregation state.[3] For PSBr we initially measured the depolarization behavior at a scattering angle of 90°. This is shown in Figure 2, where the distribution of $C^v(R_{90})$ *vs.* molar mass has been overlaid upon the MMD of PSBr.[3] We subsequently measured the relationship between the Cabannes factor, the scattering angle (θ), and the molar mass as a continuous function of the latter for a multiplicity of values of θ.[18] Values of $C^v(R_\theta)$ were observed to increase with decreasing molar mass, a result postulated to arise from the restricted configurational entropy imposed by the heavy atom substituent. It was also noted that, regardless of molar mass, $C^v(R_\theta)$ was always smallest at $\theta = 90°$.

During studies of poly(γ-benzyl-L-glutamate), PBLG, using SEC with depolarized right-angle light scattering,[3] we were initially surprised both by the small change in $C^v(R_{90})$ as a function of molar mass and with how close to unity $C^v(R_{90})$ was. Our inability to observe significant depolarization in solutions of this molecule was principally ascribed to a combination of structural factors and experimental set-up, and secondarily to thermodynamic and geometric (angular) conditions. Flow rate was later considered a possible factor as well. Regarding the purported principal cause: It has been noted that in non-aggregating solvents (*e.g.*, DMF) PBLG retains an α-helix conformation and in dilute solution bears more resemblance to the "wormlike chain" model of Kratky and Porod than to the "rigid rod" model of Kuhn.[15,23] Originally referred to as a rigid rod, PBLG is commonly referred to nowadays as a *semi-flexible polymer*, i.e., as being stiffer than a flexible polymer but not quite a rigid rod.[15] We see evidence of this under the present experimental conditions in the conformation plot (Figure 3), which shows the double-logarithmic relationship between the root-mean-square radius of the polypeptide and its molar mass. The slope of this plot, α, is 0.93, corresponding to a fractal dimension ($\alpha \equiv 1/d_f$) for PBLG of 1.08, slightly above that of a rigid rod (for which $d_f = 1$) but well below that of a random coil in either a good ($1.67 \le d_f < 2$) or a theta ($d_f = 2$) solvent.[2,24] Using the MALS detector, no evidence of aggregation was observed for PBLG in DMAc/LiCl, as noted by the lack of extraneous peaks or shoulders in the LS signals at all seventeen angles of observation as well as by the constancy in the slope of the conformation plot.

$C^v(R_{90})$ for PBLG did increase steadily with decreasing molar mass, but only minimally, from 1.006 at 297 Kg/mol to 1.018 at 78 Kg/mol. We postulated that the observed lack of depolarization was due to the fact that the design of the flow cell in the MALS photometer is such that the flow path is aligned in a parallel, line-of-sight geometry with the optical path of the laser. In its passage through the flow cell PBLG, a polymer with limited flexibility, was presumed to align with the streamlines of flow. Unfortunately, experimental values of the rotary diffusion coefficient of PBLG[4] are about two orders of magnitude lower than typical shear rates experienced by macromolecules in the flow cell of a light scattering photometer[25] and, thus, the purported flow-induced

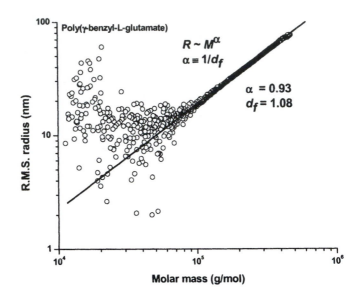

Figure 3. Conformation plot (log root-mean-square radius vs. log molar mass) of PBLG, as determined by SEC/MALS. Open circles correspond to experimental data, solid line to linear fit of data between 70000 and 470000 g/mol.

alignment is extremely unlikely at the flow rates common to most SEC experiments.

We also noted that it might be possible to observe depolarization from solutions of PBLG in a different solvent or at a higher concentration. With respect to the former, an extremely high segmental anisotropy has been measured for PBLG in dichloroethane, approximately two orders of magnitude higher than the value measured in dichloroacetic acid (both values include a form effect due to the non-matching nature of the solvents).[26] In the former solvent the molecule retains its helical configuration, while in the latter it adopts a coiled structure. The effect of the solvent on the depolarization ratio of polymers was noted previously by Picot *et al.*[27] We also believe that we are currently working below the critical concentration at which PBLG will phase-separate into two phases, one of these being a nematic liquid crystalline phase. In dioxane, this critical concentration appears to occur at a volume fraction of 0.058,[28] but a higher critical concentration is expected in DMF due to PBLG displaying a higher order in dioxane than in the amide solvent.[29] It is reasonable to extrapolate this expectation to the case of DMAc/LiCl.

While we did not undertake the measurement of $C''(R_\theta)$ in different solvents or at different flow rates, we did investigate the angular dependence of the Cabannes factor for PBLG in DMAc/LiCl using SEC/D-MALS. Figure 4 shows the ternary relationship between Cabannes factor, scattering angle, and molar mass as a continuous function of the latter. The large deviations in both positive and negative direction at the extremes of the plots are due to the low signal-to-noise ratios (S/N) in these regions as well as, in the low molar mass region, to the strong depolarization behavior of oligomers. To assist in interpreting the information in Figure 4, we graphed the Cabannes factors for two molar mass slices in Figure 5. These slices are near the extremes of the "Cabannes factor gradient," i.e., $\Delta C''(R_\theta)$ is greatest between them if one does not consider the regions with both positive and negative deviations just mentioned. It may be observed that, for both molar mass slices, the Cabannes factor decreases steadily as a function of scattering angle until $\theta = 90°$, subsequent to which the Cabannes factors show a rapid increase (similar behavior was noted for PSBr).[18] The lack of significant depolarization, though, occurs irrespective of scattering angle. The largest depolarization gradient as a function of molar mass appears to be at $\theta = 72°$, though it is still quite small at this angle. The molar mass dependence of the Cabannes factor, previously observed for other types of polymers, is also seen to have an angular dependence as seen for PBLG in Figures 4 and 5.

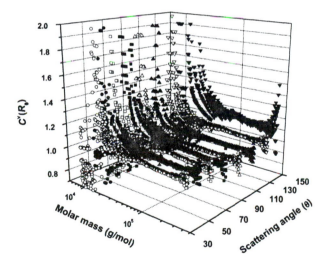

Figure 4. Distribution of the Cabannes factor of PBLG as a function of molar mass as a function of scattering angle. Detection angles are, from lowest to highest, 30°, 43°, 56°, 72°, 90°, 108°, 127°, 142°.

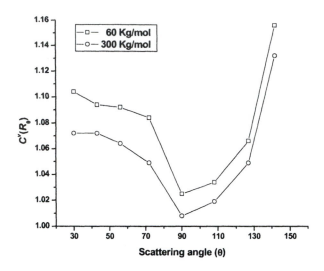

Figure 5. Cabannes factors for two molar mass slices of PBLG (60000 and 300000 g/mol) as a function of scattering angle. Data are from same detection angles as in Figure 4. Solid lines are placed on graph to guide the eye and are not meant to imply continuity between data points.

Detection of impurities in a sample

A different application of SEC/D-MALS is in the detection of small amounts of impurity in a polymeric sample, where the impurity may depolarize light in a manner different from that of the sample. In Figure 6 we have overlaid the right-angle scattering chromatograms for a particular polymer, termed here Sample A. Separation by means of size-exclusion chromatography allows discrimination between the large analyte and the small impurity based on elution volume/time. The smooth solid line is the SEC/MALS signal (i.e., without Polaroid filters), the dashed line is the SEC/D-MALS signal with the filters in vertical position (i.e., with the transmission axis of the filters parallel to the polarization axis of the laser), and the jagged solid line is the SEC/D-MALS signal with the filters in the horizontal position (transmission axis perpendicular to the polarization axis of the laser). Sample A has a retention time of ~25 minutes, the impurity a retention time of ~35 minutes. The impurity is barely distinguishable from the baseline when the sample was analyzed both without the filters and in vertical polarization mode. However, when the sample was analyzed in horizontal polarization mode the impurity manifested itself quite

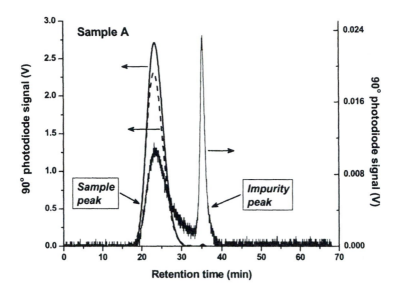

Figure 6. 90° photodiode data of Sample A: SEC/MALS (smooth solid line), SEC/D-MALS in aligned-polarization mode (dashed line), SEC/D-MALS in cross-polarization mode (jagged solid line).

clearly. While the scales of the different y-axes in the figure obviously differ, and while small yet noticeable peaks can be seen in the SEC/MALS and vertical-polarization (aligned mode) SEC/D-MALS traces, the S/N for the impurity peak in the former is ~2:1 and less than that in the latter. In the horizontal-polarization (crossed mode) SEC/D-MALS experiment the S/N for the same component is >60:1.

Batch-mode D-MALS[11,31]

Study of the dependence of the mean-square optical anisotropy, $\langle \Gamma^2 \rangle$, on molar mass has the ability to provide useful information about chain stiffness and local chain conformations.[32] The mean-square optical anisotropy of polymers may be calculated by, *e.g.,* equation (8):

$$\left\langle \Gamma^2 \right\rangle = \frac{15\lambda_0^{\,4}}{16\pi^4} \frac{M_w}{N_A} \left(\frac{3}{n_0 + 2} \right)^2 \left(\frac{\Delta R_h^v}{c} \right)_{c=0,\theta=0} \qquad (8)$$

where λ_0 is the wavelength of the incident radiation, N_A is Avogadro's number, n_0 is the refractive index of the solvent, and c is the concentration of the analyte in solution. For the last term on the right side of equation (8):

$$\lim_{c,\theta \to 0} \frac{\Delta R_h^v}{2Kc} = 3\delta M_w \qquad (9)$$

where K is the familiar optical constant equal to $4\pi^2 n_0^2 (\partial n/\partial c)^2 \lambda_0^{-4} N_A^{-1}$ and δ is a parameter representing the anisotropy of the local polarizability tensor affixed to the statistical segment (or to the chain contour). The same extrapolation that yields M_w when plotting the data from a batch-mode MALS experiment in Berry square-root form yields δ when plotting the data from a batch-mode crossed-polarizer D-MALS experiment in the same form. We note that, to our awareness, all batch-mode D-MALS data in the literature to date have been acquired with variable-angle instruments, not with photometers that detect the light scattered at a multiplicity of angles simultaneously. The latter type is, of course, what we have utilized as part of our SEC/D-MALS set-up and which can be taken off-line quite easily for batch-mode experiments.

Determining δ allows for optical anisotropy corrections to M_w and $\langle r^2 \rangle^{1/2}$ as determined by the Berry square-root plot of $(2Kc/\Delta R_u^v)^{1/2}$ vs. c and/or $\sin^2(\theta/2)$, as the molar mass determined from this plot is not M_w but an apparent weight-average molar mass $M_{w,app} = M_w(1 + 7\delta)$.[31] Nakatsuji et al. have recently applied the procedure described in this sub-section to the semi-flexible polymer poly(n-hexyl isocyanate).[31] They determined the mean-square optical anisotropy of this polymer as a discrete function of its degree of polymerization by studying nine narrow polydispersity samples, ranging in molar mass from 16500 g/mol to 104000 g/mol, in n-hexane at 25 °C. The behavior of the ratio $\langle \Gamma^2 \rangle / x_w$ as a function of the weight-average degree of polymerization x_w was explained by the theory on the basis of the Kratky-Porod wormlike chain model.

Conclusions

SEC/D-MALS bears a superficial resemblance to rheo-optical methods, but upon closer inspection these techniques are seen to provide quite different types of information. Techniques such as flow birefringence are used to study extensional flows, to visualize stress fields, to study the dynamics of homo- and block copolymers and of copolymer blends, etc. The coupling of size-exclusion chromatography to depolarized multi-angle light scattering gives information critical for accurate calculations of molar mass via both batch-mode and flow-through light scattering, measures the depolarization of dilute polymer solutions as a function of observation angle, and presents us with a new way to determine impurities in a polymeric sample. It also provides a means of studying the form and stiffness of polymers in solution, specially of macromolecules that become more rigid as a function of decreasing molar mass.

Experiments of this type are currently not easy to perform. One can foresee, though, the incorporation of a (low cost!) "drop-down" set of filters into a MALS photometer such that manipulation of a simple lever or the touch of a button would bring the filters into position, to then be removed just as easily. This set-up would probably go a long way toward popularizing D-MALS experiments as well as rendering them suitable to the study of polymer solutions with time-dependent properties. Coupling to other types of separation method such as field-flow fractionation, hydrodynamic chromatography, etc., in the form of FFF/D-MALS or HDC/D-MALS has the potential to relate the depolarization data to a variety of fundamental polymer properties.

Acknowledgements

The author would like to thank Dr. David B. Alward (Surface Specialties UCB), Prof. Shelley L. Anna (Dept. of Mechanical Engineering, Carnegie Mellon University), Dr. Patricia M. Cotts (Dupont CR&D) and Prof. Paul S. Russo (Dept. of Chemistry, Louisiana State University) for many helpful discussions and support during the course of these studies.

References

1. (a) de Gennes, P.-G. *The Physics of Liquid Crystals;* Clarendon Press: Oxford, 1974. (b) Chandrasekhar, S. *Liquid Crystals;* Cambridge University Press: Cambridge, 1977.
2. Burchard, W. *Adv. Polym. Sci.* **1999,** *143,* 113-194.
3. Striegel, A. M. *Anal. Chem.* **2002,** *74,* 3013-3018.
4. Tanford, C. *Physical Chemistry of Macromolecules;* Wiley: New York, 1961.
5. (a) Hecht, E. *Optics, 2nd edition;* Addison-Wesley; Reading, MA, 1987. (b) Shurcliff, W. A. *Polarized Light –Production and Use*; Harvard University Press; Cambridge MA, 1962.
6. (a) Hiemenz, P. C. *Polymer Chemistry;* Marcel Dekker: New York, 1984. (b) Katime, I. A.; Quintana, J. R. In *Comprehensive Polymer Science,* Booth, C., Price, C., Eds.; Pergamon: Oxford, 1989; Vol. 1, pp 103-132. (c) van de Hulst, H. C. *Light Scattering by Small Particles;* Dover: New York, 1957. (d) Kliger, D. S.; Lewis, J. W.; Randall, C. E. *Polarized Light in Optics and Spectroscopy;* Academic Press: Boston, 1990.

7, Cabannes, J.; Rocard, M. Y. *La Diffusion Moléculaire de la Lumière;* Les Presses Universitaires de France: Paris, 1929.

8. (a) Martin, W. H. *Trans. Royal Soc. Canada,* **1923,** *III,* 151-155. (b) Kerker, M. *The Scattering of Light and Other Electromagnetic Radiation;* Academic Press: New York, 1969.

9. Chu, B. *Laser Light Scattering;* Academic Press: New York, 1974.

10. (a) Wyatt, P. J. *Anal. Chim. Acta* **1993,** *272,* 1-40. (b) Brice, B. A.; Halwer, M.; Speiser, R. *J. Opt. Soc. Am.* **1950,** *40,* 768-778.

11. Jinbo, Y.; Varichon, L.; Sato, T.; Teramoto, A. *J. Chem. Phys.* **1998,** *109,* 8081-8086.

12. Berne, B. J.; Pecora, R. *Dynamic Light Scattering;* Dover: Mineola, NY, 1976.

13. Macosko, C. W. *Rheology –Principles, Measurements, and Applications;* VCH Publishers: New York, 1994.

14. Fuller, G. G. *Optical Rheometry of Complex Fluids;* Oxford University Press: New York, 1995.

15. Larson, R. G. *The Structure and Rheology of Complex Fluids;* Oxford University Press: New York, 1999.

16. Miller, M. L. *The Structure of Polymers;* Reinhold: New York, 1966.

17, Onuki, A.; Doi, M. *J. Chem. Phys.* **1986,** *85,* 1190-1197.

18. Striegel, A. M. *Polym. Int.* **2003,** *52,* 1863-1868.

19. Debye, P. *J. App. Phys.* **1944,** *15,* 338-342.

20. (a) Stacey, K. A. *Light-Scattering in Physical Chemistry;* Academic Press: New York, 1956. (b) Zimm, B. H.; Stein, R. S.; Doty, P. *Polym. Bull.* **1945,** *1,* 90-119, and references therein. (c) Krishnan, R. S. *Proc. Ind. Acad. Sci.* **1938,** *7A,* 98-103, and references therein. (d) Perrin, F. *J. Chem. Phys.* **1942,** *10,* 415-427. (e) Norris, F. H., Ph.D. Dissertation, University of Massachusetts, Amherst, 1956.

21. Doty, P. *J. Polym. Sci.* **1948,** *3,* 750-771.

22. Horn, P.; Benoit, H. *J. Polym. Sci.* **1953,** *10,* 29-37.

23. Schmidt, M. *Macromolecules* **1984,** *17,* 553-560.

24. (a) Mandelbrot, B. B. *The Fractal Geometry of Nature;* W. H. Freeman & Co.: New York, 1983. (b) de Gennes, P.-G. *Scaling Concepts in Polymer Physics;* Cornell University Press: Ithaca, NY, 1979.

25. Reed, W. F In *Strategies in Size Exclusion Chromatography, ACS Symposium Series 635;* Potschka, M.; Dubin, P. L., Eds.; American Chemical Society: Washington, DC, 1996; pp. 7-34.

26. Tsvetkov, V. N.; Andreeva, L. N. In *Polymer Handbook, 3rd edition;* Brandup, J.; Immergut, E. H., Eds.; Wiley: New York, 1989; pp VII/555-VII/590.

27. Picot, C.; Weill, G.; Benoit, H. *J. Colloid Interface Sci.* **1968,** *27,* 360-376.

28. Samulski, E. T. In *Liquid Crystalline Order in Polymers;* Blumstein, A., Ed.; Academic Press: New York, 1978; pp 167-190.

29. Abe, A.; Yamazaki, T. *Macromolecules* **1989,** *22,* 2145-2149.
30. Stein, R. S.; Rhodes, M. B.; Porter, R. S. *J. Colloid Interface Sci.* **1968,** *27,* 336-354.
31. Nakatsuji, M.; Ogata, Y.; Osa, M.; Yoshizaki, T.; Yamakawa, H. *Macromolecules* **2001,** *34,* 8512-8517.
32. Yamakawa, H. *Helical Wormlike Chains in Polymer Solutions;* Springer: Berlin, 1997.

Chapter 5

A Review of the Applications of Size-Exclusion Chromatography –Multiangle Light Scattering to the Characterization of Synthetic and Natural Polymers

On Experimental and Processing Parameters

Stepan Podzimek

SYNPO, 532 07 Pardubice, Czech Republic

This chapter deals with several aspects of polymer characterization by means of size-exclusion chromatography (SEC) coupled with a multi-angle light scattering (MALS) photometer. They are: (i) the determination of the refractive index (RI) detector calibration constant, (ii) the influence of the RI detector temperature and wavelength, (iii) the fit method used to extrapolate the light scattering data, (iv) the determination of the number-average molar mass, (v) the analysis of oligomers, (vi) and the analysis of polymer blends.

Review of Fundamental Principles

Combination of size-exclusion chromatography (SEC) with a multi-angle light scattering (MALS) detector has become a routine technique that brings many new possibilities to polymer analysis and characterization and solves many traditional problems of SEC, particularly the necessity for column calibration and the sensitivity of the obtained results to the flow rate variations, temperature fluctuations, column performance, and mass of injected sample.

The fundamental description of the theory, instrumentation, and various applications of SEC-MALS were reviewed in an extensive article by Wyatt (1). The intensity of light scattered by polymer molecules in dilute solution is expressed by the quantity called the excess Rayleigh ratio (R_θ) defined by the following equation:

$$R_\theta = f \frac{(I_\theta - I_{\theta,solvent})}{I_0} \qquad (1)$$

where I_θ is the scattered light intensity of the solution, $I_{\theta,solvent}$ is the scattered light intensity of the solvent, I_0 is the intensity of the incident radiation, f is an experimental constant related to the geometry of the apparatus. The subscript θ implies the angle between the scattering direction and the incident light beam.

The concentration and angular dependence of the intensity of light scattered by a dilute polymer solution can be described by one of the following equations (2,3):

$$\frac{R_\theta}{K^*c} = MP(\theta) - 2A_2cM^2P^2(\theta) \qquad (2)$$

$$\frac{K^*c}{R_\theta} = \frac{1}{MP(\theta)} + 2A_2c \qquad (3)$$

$$\sqrt{\frac{K^*c}{R_\theta}} = \frac{1}{\sqrt{MP(\theta)}} + A_2c\sqrt{MP(\theta)} \qquad (4)$$

where c is the concentration of polymer in solution (g/mL), M is the molar mass (weight average in case of polydisperse polymer), A_2 is the second virial coefficient (at the low concentrations typical of SEC separation, A_2 is usually negligible as the term $2A_2cM \ll 1$), K^* is an optical constant:

$$K^* = \frac{4\pi^2 n_0^2}{\lambda_0^4 N_A}(dn/dc)^2 \qquad (5)$$

where n_0 is the refractive index of the solvent at the incident radiation wavelength, λ_0 is the incident radiation wavelength at vacuum, N_A is Avogadro's number, dn/dc is the specific refractive index increment. Note: Equation (5) is valid for vertically polarized incident light. $P(\theta)$ is the particle scattering function (the ratio of the intensity of radiation scattered at an angle of observation θ to the intensity of radiation scattered at zero angle, R_θ /R_0) that describes the decrease of the scattered light intensity with increasing angle of observation. For small θ the particle scattering function is expressed as:

$$P(\theta) = 1 - \frac{16\pi^2}{3\lambda^2}R^2 \sin^2(\theta/2) \qquad (6)$$

where $\lambda = \lambda_0/n_0$ is the wavelength of the incident light in a given solvent, R^2 is the mean square radius (z-average in case of polydisperse polymers). The root mean square (RMS) radius (R, also called radius of gyration) describes the distribution of mass around the center of gravity and thus expresses the size of a particle regardless of its shape. Equation (6) allows the calculation of R from the initial slope of the angular variation of the scattered light intensity.

In contrast to conventional SEC, the combination of SEC with a MALS detector requires the absolute measurements of concentration at each elution volume slice. It is important to note that light scattering is undoubtedly the absolute method for molar mass determination. However, it is as absolute as the absolute values of dn/dc and of concentration that are used for data processing. The concentration c_i of the molecules eluting at the ith elution volume slice is mostly determined from the signal of a differential refractive index (RI) detector:

$$c_i = \frac{\alpha(V_i - V_{i,baseline})}{dn/dc} \qquad (7)$$

where α is the RI detector calibration constant (in RI units per volt), V_i and $V_{i,baseline}$ are the RI detector sample and baseline voltages, respectively. Alternatively, other types of concentration detectors (e.g. UV detector) can be employed for the determination of concentration.

Experimental

Various SEC-MALS set-ups consisting of an HPLC pump, a manual injector or a 717plus autosampler (Waters), styrene-divinylbenzene based SEC columns (Waters Styragel HR or Polymer Laboratories PLgel), a DAWN EOS or a miniDAWN light scattering photometer (Wyatt Technology Corporation), and an Optilab interferometric refractometer (Wyatt Technology Corporation) or a Waters 410 differential refractometer were used for the SEC-MALS measurements. The detectors operated at the wavelengths of 690 nm (MALS photometers), 633 nm or 690 nm (Optilab), and 930 nm (Waters 410). MALS detectors were calibrated by means of toluene (HPLC grade, Aldrich) using the Rayleigh ratio of 9.78×10^{-6} cm^{-1} and normalized by computing normalization coefficients for each angle using polystyrene with molar mass of 30,000 g/mol. Tetrahydrofuran (THF) inhibited with 250 ppm 2,6-di-*tert*-butyl-4-hydroxytoluene (99+ %, Aldrich) at a flow rate of 1 mL/min was used as a mobile phase. The Optilab interferometer was also used for the off-line determination of *dn/dc*. ASTRA software (Wyatt Technology Corporation) was used for the data collection and processing, DNDC software (Wyatt Technology Corporation) was employed for the Optilab calibration and *dn/dc* determination.

Influence of the Experimental and Processing Parameters on the Results Obtained by SEC-MALS

RI Detector Calibration

The determination of the RI detector calibration constant can be accomplished by injecting the solutions of a compound with accurately known *dn/dc* either directly into the RI detector (off-line mode) or in an SEC mode assuming that 100% of the injected sample elutes from the SEC columns. The former method is usually assumed to be more precise. However, based on the author's own experience, both methods provide identical results. For the sake of calibration in the SEC mode, it is worth mentioning that many manual injectors add additional volume of about 10 µL corresponding to the inside injector channels. With a typical 100 µL sample loop this results in 10% erroneous RI detector calibration constant, and consequently 10% erroneous molar masses. The true injected volume can be determined experimentally injecting a polymer with reliably known *dn/dc* that, under given conditions, elutes with 100% mass recovery. Using an autoinjector capable of injecting variable volumes instead of a manual injector eliminates this problem.

The accuracy of the RI detector calibration constant strongly depends on the accuracy of the *dn/dc* of the compound used for the detector calibration.

Aqueous solutions of sodium chloride are usually used for the off-line refractometer calibration. The *dn/dc* values of NaCl solutions at different wavelengths have been reported by Becker et al. (*4*). The experimental data in reference (*4*) cover the wavelength range of 436 – 633 nm and give the following relation between wavelength and *dn/dc* of NaCl solutions at the room temperature:

$$dn / dc = 1.673 \, x \, 10^{-7} \, \lambda_0^2 - 2.237 \, x \, 10^{-4} \, \lambda_0 + 0.2487 \qquad (8)$$

Using equation (8), a *dn/dc* of 0.174 mL/g is obtained for the two wavelengths 633 nm and 690 nm. Another *dn/dc* of aqueous NaCl of 0.179 mL/g can be found for 633 nm in reference (*5*).

Polystyrene (PS) well satisfies the requirement of 100% mass recovery from the columns and can be recommended for RI detector calibration in SEC mode with THF as the mobile phase. Values of *dn/dc* in the range of 0.180 – 0.193 mL/g have been reported in the literature for PS in THF for the wavelength range 633 – 670 nm and the temperature range 20 – 40 °C (*6*). The values of *dn/dc* of PS (THF, room temperature, 633 nm) determined by the author using the Optilab RI are as follows: 0.183 ± 0.001 mL/g when the *dn/dc* for NaCl of 0.174 mL/g was used for the Optilab calibration, and 0.188 ± 0.001 mL/g when a *dn/dc* for NaCl of 0.179 mL/g was employed. Therefore, the value of 0.185 ± 0.001 mL/g can be considered as a reliable approximation of the *dn/dc* of PS in THF in the wavelength range 633 – 690 nm.

Determination of the Specific Refractive Index Increment

Knowing the *dn/dc* value is absolutely essential to the determination of molar mass by light scattering. Even though values of *dn/dc* have been published for various polymers, the choice of a reliable figure may not always be obvious. If a trustworthy literature *dn/dc* value for the analyzed polymer is unavailable, it must be determined experimentally using exactly the same procedure as that employed for the RI detector calibration (i.e. using the RI detector off-line or in SEC run assuming 100% mass recovery of injected sample). The uncertainty of the experimentally obtained *dn/dc* is given solely by the uncertainty in the RI detector calibration constant.

Purity of the compound used for the RI detector calibration or the *dn/dc* determination is crucial to obtain accurate results. It should be emphasized that some hygroscopic compounds such as NaCl or dextran can contain significant amounts of water.

Influence of RI Detector Temperature and Wavelength on Molar Mass Data

A potentially important issue concerning the determination of concentration is the temperature dependence of both dn/dc and the RI detector calibration constant. In order to obtain a stable signal the operating temperature of an RI detector is often set up 10 - 15 °C above room temperature while the literature dn/dc values are mostly available for room temperature. Table I lists the weight-average molar masses (M_w) obtained for a PS sample at three different temperatures using dn/dc and RI calibration constants valid for room temperature. The data show no significant influence of the RI detector temperature on the obtained results within the investigated temperature range.

Table I. Influence of the RI detector temperature on M_w of PS

RI detector temperature (°C)	M_w (10^3 g/mol)
23	208.9 ± 0.2
35	210.3 ± 0.3
50	206.5 ± 0.4

M_w results are averages of 4 - 6 measurements.

Because of the wavelength dependence of dn/dc, the dn/dc value at the wavelength of the MALS detector should be used for the calculation of molar mass. However, many RI detectors operate at the wavelength different from that of the MALS detector. In such case the optical constant K^* is calculated using a dn/dc valid for the wavelength of the MALS photometer while the concentration is determined using the equation (7) with a dn/dc that does not match the operating wavelength of the RI detector. The possible error arising from this fact is investigated in Table II where M_w of several different polymers obtained by means of the Optilab operating at 690 nm are compared with those obtained with the 410 RI detector operating at 930 nm. Both RI detectors were calibrated in the SEC mode by multiple injections of known amounts of PS using the dn/dc of PS at 690 nm. The calculations of M_w were carried out using dn/dc values at 690 nm. The results obtained by both RI detectors compare well to each other. This finding can have two causes: (i) the differences in the dn/dc values of polymers under investigation at the wavelengths of 690 nm and 930 nm are negligible; (ii) the wavelength dependence of the dn/dc of PS and the analyzed polymers are analogous. As a result of the latter, the error in the RI detector calibration constant (caused by erroneous dn/dc of PS used for the calibration) is compensated by the erroneous *dn/dc* values of the particular polymers and the concentrations calculated using equation (7) are correct. However, experimental evidence supporting either of the two hypotheses or their general validity is not available. Therefore, an RI detector operating at the wavelength of the MALS detector certainly provides more trustworthy results.

Table II. Comparison of M_w determined by SEC-MALS using Optilab at 690 nm and 410 RI detector at 930 nm

Polymer	M_w (10^3 g/mol)	
	Optilab	410 detector
EPBA	8.3	8.4
PS	90	89
PBZMA	426	423
PIB	430	424
PMMA	570	559

EPBA = epoxy resin based on bisphenol A, PS = polystyrene, PBZMA = poly(benzyl methacrylate), PIB = polyisobutylene, PMMA = poly(methyl methacrylate).

Results uncertainty < ± 5 %.

Fit Method

According to equations (2) – (4), a plot of either R_θ/K^*c, K^*c/R_θ or $(K^*c/R_\theta)^{1/2}$ versus $\sin^2(\theta/2)$ yields a curve whose intercept and slope at zero angle are used for the calculation of molar mass and RMS radius, respectively. Equations (2) – (4) describe the same phenomenon by means of slightly different formalisms. Concordant with ASTRA software, the terms Debye, Zimm, and Berry fit methods will be employed henceforth for equations (2) – (4), respectively. Additionally, the so called Random coil fit method uses the theoretically derived particle scattering function (7) instead of fitting a polynomial to $\sin^2(\theta/2)$. Tables III and IV compare the molar masses and RMS radii calculated using the various fitting methods. These data lead to the following conclusions:

1. All formalisms provide similar results.
2. The Debye fit method requires different polynomial degrees to fit the light scattering data, and therefore it may not be suitable for processing samples with broad molar mass distribution, because the linear extrapolation of

equation (2) is insufficient to fit the data for higher molar masses while higher order fits are too high for smaller molecules. A high degree polynomial results in excessive freedom in the extrapolation to zero angle, leading to incorrect molar mass and RMS radius.

3. The great advantage of the Zimm fit method over the Debye method is linearity with respect to $\sin^2(\theta/2)$ up to high molar masses. Compared to other formalisms the method slightly overestimates molar masses and more considerably RMS radii of large molecules.

4. The selection of an appropriate fit method becomes important when branching is of interest as can be illustrated with data of one of the randomly branched PS in Tables III and IV, for which the branching ratio of 0.41 is derived using molar mass and RMS radius obtained by the Zimm method while the branching ratios of 0.30, 0.32, and 0.34 are derived using the Debye, Berry and Random coil methods, respectively. The Zimm formalism generally yields slightly higher slopes of the RMS radius versus molar mass plots compared to those obtained by the Berry or Random coil methods.

5. The Berry fit method is linear over a broad molar mass range and can be a good alternative for samples with very broad molar mass distribution.

6. The Random coil method provides results comparable to the Debye and Berry formalisms even in case of compact, branched macromolecules.

Determination of the Number-Average Molar Mass

Knowledge of the number-average molar mass (M_n) is necessary for evaluating polymerization kinetics, determination of end groups, or stoichiometric calculations. The glass transition temperature of a polymer is particularly sensitive to M_n. The ratio of M_w/M_n is a measure of the breadth of the molar mass distribution. Therefore, M_n is a characteristic of prime importance to a polymer sample. Conventional SEC, end-group analysis (EG), membrane osmometry (MO), and vapor pressure osmometry (VPO) are regularly used for the determination of M_n. The significant weakness of SEC is that it is a relative technique and that careful calibration for each polymer type is required. VPO and EG techniques are applicable solely for polymers with $M_n < 20,000$ g/mol, as erroneous MO results (overestimation of M_n) are obtained when a portion of the sample permeates through the membrane. Therefore, MO is not suitable for polydisperse samples containing oligomeric fractions. In addition, the absolute methods require exact knowledge of solution concentrations, the osmometric measurements have to be carried out at multiplicity of concentrations, low molar mass impurities (residual raw materials, solvents, or moisture) severely decrease the M_n determined by VPO, and EG results may be affected by the change of end groups due to side reactions or the presence of cyclic or branched molecules.

Table III. Comparison of molar masses calculated by various light scattering formalisms for linear and branched molecules

Polymer	M_w (10^3 g/mol)			
	Zimm	Debye	Berry	Random coil
Linear PS	210	211	211	211
Linear PS	586	589[*]	586	588
Linear PS	949	949[*]	945	948
Linear PS	1620	1610[**]	1598	1603
Linear PS	2148	2120[**]	2100	2111
Linear PS	3930	3753[**]	3821[*]	3770
Linear PS	6073	5853[***]	5893[*]	5753
Randomly branched PS	3147	3102[**]	3100	3107
Randomly branched PS	7836	7453[**]	7496	7583
Star branched PBZMA	3700	3695	3699	3699

[*]Second, [**]third, and [***]fifth order polynomial was used to fit the angular variation of scattered light intensity. First order fit was used for unlabeled data.

Note: Molar masses were calculated for very narrow peak sections.

Table IV. Comparison of RMS radii calculated by various light scattering formalisms for linear and branched molecules

Polymer	R_z (nm)			
	Zimm	Debye	Berry	Random coil
Linear PS	17.8	17.3	17.7	17.7
Linear PS	33.1	32.7[*]	32.3	32.4
Linear PS	44.7	43.0[*]	42.8	42.9
Linear PS	63.3	60.3[**]	58.5	59.0
Linear PS	76.5	70.6[**]	68.9	69.9
Linear PS	117.2	98.1[**]	107.0[*]	102.3
Linear PS	149.4	130.8[***]	136.0[*]	128.0
Randomly branched PS	63.8	58.4[**]	58.6	59.2
Randomly branched PS	97.0	80.6[**]	83.6	86.3
Star branched PBZMA	21.8	20.7	21.5	21.5

[*]Second, [**]third, and [***]fifth order polynomial was used to fit the angular variation of scattered light intensity. First order fit was used for unlabeled data.

Note: RMS radii were calculated for very narrow peak sections.

In case of SEC-MALS, M_w is measured by first principles and does not depend of the SEC separation, while M_n is calculated assuming that molecules eluting at particular elution volume slices are monodisperse with respect to molar mass. However, due to the limited resolution of SEC columns, particular volume slices are not absolutely monodisperse and the molar mass measured by the MALS detector at each elution volume slice is actually the weight average.

Consequently, the SEC-MALS method tends to overestimate M_n values and the results might be expected to be strongly dependent on SEC resolution. The effect of SEC resolution on M_n obtained by SEC-MALS is demonstrated in Figure 1 and Table V. Surprisingly, there is no obvious effect of increasing resolution on the obtained M_n. The ability of SEC-MALS to provide reliable M_n is further demonstrated on the well-known polydisperse PS standard NIST 706 (previously labeled as NBS 706) with the originally reported value of $M_n = 137,000$ g/mol (measured by MO). The RI chromatogram of NIST 706 (Figure 2) is flat and parallel with the baseline at the region of high elution volumes. Similar chromatogram patterns are often encountered with technical polymers. The reason for this incomplete baseline recovery in the present case is the existence of oligomeric fractions as proved in reference (8). Also, non-SEC effects can result in a non-returning RI baseline. Selection of the cutoff elution

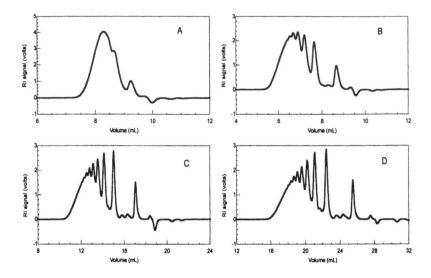

Figure 1. Effect of SEC resolution on the number-average molar mass. RI chromatograms of epoxy resin obtained using different SEC column sets, A: 1 x PLgel Mixed-C 5 μm 300 x 7.5 mm, B: 1 x PLgel Mixed-E 3 μm 300 x 7.5 mm, C: 2 x PLgel Mixed-E 3 μm 300 x 7.5 mm, D: 3 x PLgel Mixed-E 3 μm 300 x 7.5 mm; M_n = 1,650 ± 10; 1,610 ± 30; 1,600 ± 50; 1,600 ± 60 g/mol for sets A, B, C, and D, respectively. The results are averages of 6 – 9 measurements.

volume may represent a serious concern for processing the polymer samples of this type.

There are four different cutoff elution volumes equivalent to different molar masses marked on the RI chromatogram in Figure 2. The corresponding results (Table VI) demonstrate strong dependence of M_n on selected cutoff elution volume. Inclusion of molar masses down to a few hundred g/mol gives an M_n much below the nominal value, while the cutoff at 10,000 g/mol provides an M_n close to the nominal value. Because the higher M_n obtained by MO can be explained by permeation of oligomeric molecules through the membrane, the lower M_n measured by SEC-MALS is a more reliable result.

In view of the fact that oligomeric fractions may influence polymer properties significantly, all chromatographic data up to the beginning of the solvent peak should be included in the calculation of M_n. In the case of polydisperse polymers containing a substantial amount of oligomeric fractions, some error may arise from the non-constancy of dn/dc in the oligomeric region.

Table V. Effect of resolution on M_n of broad PS ($M_w/M_n = 2.8$)

No. of SEC columns	R_{SP}	M_n (g/mol)
1	2.2	118 ± 1
3	3.1	121 ± 5

The specific resolution $R_{SP} = \dfrac{2(V_2 - V_1)}{(w_1 + w_2)\log(M_1/M_2)}$, where V is the elution volume of narrow PS standard of molar mass M, and w is the baseline peak width. PS standards of M 68,000 g/mol and 675,000 g/mol were used for the determination of R_{SP}.

M_n values are averages of 8 measurements.

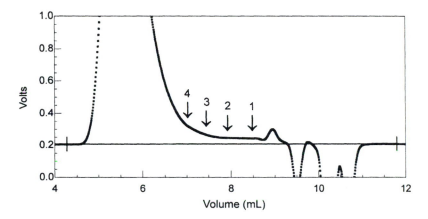

Figure 2. RI chromatogram of PS NIST 706 showing four different cutoff elution volumes.

SEC-MALS Analysis of Oligomers

Oligomers such as various types of synthetic resins or natural oligosacharides find utilization in various fields. VPO and EG are the absolute methods of choice for measuring M_n of oligomers. The limitations of these methods have been discussed in the previous section.

Historically, the light scattering technique was not considered suitable for characterization of oligomers. Naturally, the low molar mass of oligomers results in generally lower sensitivity of the light scattering detectors for these materials. However, in contrast to the high molar mass polymers for which the concentrations of injected solutions less than 0.2% w/v are often required in order to eliminate undesirable column overloading and viscous fingering effects, solutions of oligomers in the range of a few percent can be injected into SEC columns without significant effect on separation performance, and thus the effect of low molar mass can be compensated.

Table VI. Number-average molar masses of PS NIST 706 calculated using four different cutoff elution volumes (see Figure 2)

Position	Cutoff M (g/mol)	M_n (10^3 g/mol)
1	550	49 ± 1
2	2,000	88 ± 1
3	5,000	112 ± 1
4	10,000	130 ± 1

The results are averages of 9 measurements.

Another issue concerning the SEC-MALS analysis of oligomers is the molar mass dependence of dn/dc that is remarkable exactly in the region of low molar masses. The different electronic environment of the end groups from that of the polymeric chain is the reason for the molar mass dependency of dn/dc since, with decreasing molar mass, the end groups represent an increasingly substantial portion of the polymer molecule. The change of dn/dc with molar mass influences both the determination of the concentration (equation (7)) and the calculation of the molar mass (equations (2) – (4)). An example of the change of dn/dc with molar mass is demonstrated in Figure 3 on PS and epoxy resin oligomers. The data in Figure 3 make obvious that dn/dc varies with molar mass within only a few thousands of g/mol.

An oligomeric sample with a broad molar mass distribution consists of molecular species with different dn/dc. If an average dn/dc value for the entire sample is used for data processing, the dn/dc is overestimated for lower molar masses and underestimated for the higher molar masses. Consequently, the molar mass and concentration are underestimated in the low molar mass region, while in the region of higher molar masses both quantities are overestimated. An error of E% in dn/dc results in the same percent error in the derived molar mass.

Table VII contrasts M_n values determined by SEC-MALS for various oligomers with those measured by means of VPO, EG, or high performance liquid chromatography (HPLC). The data do not show any obvious effect of

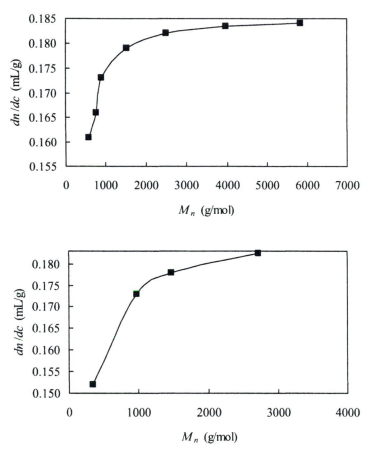

Figure 3. Dependence of dn/dc on M_n for oligomers of PS (top) and epoxy resin based on bisphenol A (bottom). The upper borders of the graphs correspond to the dn/dc of high molar mass PS and epoxy resin.
Data uncertainty < ± 0.002 mL/g.

polydispersity, although the influence of the change of dn/dc with molar mass might be expected to be more pronounced for polydisperse samples. This suggests a mutual compensation of the counteracting factors in the calculation of M_n (molar masses of lower molar mass fractions are underestimated, but they are less counted as their concentrations are underestimated as well).

It can be concluded that SEC-MALS provides reliable M_n values of oligomeric materials within 10% range of the results measured by reference techniques, and that in most practical cases the molar mass dependence of dn/dc need not be taken into account.

Polymer Blends and Copolymers

Copolymer molecules can show heterogeneity of chemical composition in addition to the distribution of molar mass. In the case of copolymers, the hydrodynamic volume depends not only on the molar mass but also on the chemical composition. Consequently, molecules of different molar mass and composition can elute at the same elution volume. Therefore, increased non-uniformity of elution volume slices with respect to both molar mass and chemical composition has to be expected in the SEC separation of copolymers as compared to homopolymers. Classical light scattering measurements of copolymers yield apparent molar masses (10-12). The influence of the chemical heterogeneity of copolymers on the results obtained by SEC with a light scattering detector has been studied theoretically (13). The study showed that SEC with a light scattering detector yields a good approximation of the molar mass distribution of a copolymer sample, provided the difference between the specific refractive index increments of the parent homopolymers is not extremely large.

Table VIII presents calculated and experimental molar mass averages for blends of three homopolymers, namely PS, poly(benzyl methacrylate) (PBZMA), and poly(methyl methacrylate) (PMMA) having dn/dc of 0.185, 0.144, and 0.084 mL/g, respectively. The blends were prepared by mixing the particular homopolymer solutions in various ratios. During SEC separation each slice contained two (three) different polymers, the mutual ratio of which varied with elution volume. The molar masses of each homopolymer within each elution slice were different. The molar mass averages listed in Table VIII were determined using an average dn/dc for the polymer blend calculated according to the equation:

$$dn/dc = \sum_i (dn/dc)_i w_i \qquad (9)$$

where $(dn/dc)_i$ and w_i are the particular homopolymer-specific refractive index increment and weight fraction in the blend, respectively.

Table VII. M_n values determined by SEC-MALS versus results from VPO, NMR or HPLC

Compound	M_n (g/mol)		M_w/M_n
	$VPO / NMR^+ / HPLC^{++}$	SEC-MALS	
PS	580[+]	630	1.1
PS	760[+]	830	1.1
PS	1,300	1,420	< 1.1
PS	1,670	1,840	< 1.1
PS	3,960	4,160	< 1.1
PS	5,020	5,230	< 1.1
PS	5,830	5,940	< 1.1
DGEBA[*]	340	370	1.0
EPBA	1,470	1,620	2.1
EPBA	2,710	2,900	2.8
DHDPM[*]	200	214	1.0
PF	420	460	2.5
PF	510	560	4.2
PBD	1,510	1,650	< 1.1
PBD	4,930	4,810	< 1.1
PTHF[**]	940[++]	900	1.7
MA-MMA	3,000	3,100	2.2

[*]Pure compound (molar mass calculated from summary formula).

[**]HPLC results for this sample reported in reference (9).

PS = polystyrene, DGEBA = diglycidyl ether of bisphenol A, EPBA = epoxy resin based on bisphenol A, DHDPM = dihydroxy diphenyl methane, PF = phenol formaldehyde resin, PBD = polybutadiene, PTHF = polytetrahydrofuran, MA-MMA = copolymer of methyl acrylate and methyl methacrylate.

M_n uncertainty $< \pm 10$ %.

Table VIII. Comparison of calculated and experimental M_n and M_w for blends of PS, PMMA, and PBZMA

Blend composition (weight fraction)	M_n $(10^3$ g/mol)		M_w $(10^3$ g/mol)	
	Calc.	Exp.	Calc.	Exp.
100 PS	-	120	-	325
100 PMMA	-	200	-	555
100 PBZMA	-	207	-	418
0.25 PS / 0.75 PMMA	171	181	498	473
0.5 PS / 0.5 PMMA	150	160	440	410
0.74 PS / 0.26 PMMA	134	145	385	364
0.25 PS / 0.75 PBZMA	175	180	395	392
0.5 PS / 0.5 PBZMA	152	150	372	367
0.74 PS / 0.26 PBZMA	135	136	349	345
0.25 PMMA / 0.75 PBZMA	205	207	452	451
0.52 PMMA / 0.48 PBZMA	203	197	489	487
0.74 PMMA / 0.26 PBZMA	202	223	519	518
0.33 PS / 0.33 PMMA / 0.34 PBZMA	165	169	433	416

Satisfactory agreements of experimental and theoretical M_n and M_w verify the theoretical conclusions stated in reference (*11*). Because homopolymer blends represent systems of maximum chemical heterogeneity, even more favorable results can be expected in the case of less heterogeneous copolymers.

Conclusions

1. A *dn/dc* value for PS in THF of 0.185 mL/g can be recommended for the calibration of RI detectors in SEC mode.
2. Regarding the calibration of the RI detector in SEC, one should keep in mind that manual injectors may add a substantial additional volume to the sample loop.
3. RI detector temperature has little effect on the obtained molar masses.
4. The use of an RI detector operating at a wavelength different from that of the MALS detector does not lead to relevant errors in the calculated molar masses.
5. Different light scattering formalisms provide similar results. The linearity of the Zimm formalism is an advantage, simplifying processing of experimental data in comparison with the Debye formalism that requires different polynomial fit degrees for various molar masses. However, the Zimm method may not be suitable for high molar mass polymers especially if the characterization of branching is an issue. In such cases the Berry or Random coil fit methods should be applied.
6. Cutoff elution volume is obviously a more serious parameter influencing M_n determination than is SEC resolution. A single, mixed pore size SEC column proved adequate for the determination of M_n. This finding is favorable owing to the general desire to reduce both time and the volumes of organic solvents required for the measurements. In the case of polymers containing oligomeric fractions, SEC-MALS provides more accurate M_n than traditionally used MO.
7. SEC-MALS is a reliable technique for the determination of M_n of oligomers. The deviation in M_n values determined by SEC-MALS from those obtained by absolute methods is within a 10% range. The molar mass dependence of *dn/dc* does not represent a serious obstacle to the characterization of oligomers by SEC-MALS.
8. Molar mass averages close to correct values can be estimated even in case of chemically heterogeneous polymer blends.

Acknowledgement

Many of the presented data and experience were collected during the author's stay with Wyatt Technology Corporation in Santa Barbara, CA. The author would like to thank to Dr. Philip Wyatt, CEO of Wyatt Technology Corporation, for the kind invitation to his company, many valuable comments, and inspirational discussions.

References

1. Wyatt, P. J. *Anal. Chim. Acta* **1993**, *272*, 1 - 40.
2. Zimm, B. H. *J. Chem. Phys.* **1948**, *16*, 1093 - 1099.
3. Berry, G. C. *J. Chem. Phys.* **1966**, *44*, 4550.
4. Becker, A.; Kohler, W; Muller, B. *Berichte der Bunsengesellschaft fur Physikalische Chemie* **1995**, *99*, 600 - 608. (In German)
5. Huglin, M. B. In *Polymer Handbook*, 3rd ed; Brandrup, J., Immergut, E. H., Eds.; John Wiley & Sons: New York, 1989, p VII/411.
6. Mori, S.; Barth, H. G. *Size Exclusion Chromatography*, Springer: Berlin, Germany, 1999, p 217.
7. Debye, P. *J. Phys. Coll. Chem.* **1947**, *51*, 18.
8. Mori, S.; Kato, H.; Nishimura, Y. *J. Liq. Chrom. & Rel. Technol.* **1996**, *19*, 2077 - 2087.
9. Rissler, K; Socher, G.; Glockner, G. *Chromatographia* **2001**, *54*, 141 - 150.
10. Benoit, H.; Froelich, D. In *Light Scattering from Polymer Solutions*, Huglin, M. B., Ed.; Academic Press: London, 1972, p 470.
11. Stockmayer, W. H.; Moore, L. D.; Fixman, M.; Epstein, B. N. *J. Polymer Sci.* **1955**, *16*, 517 – 530.
12. Bushuk, W.; Benoit, H. *Canad. J. Chem.* **1958**, *36*, 1616 – 1626.
13. Netopilik, M.; Bohdanecky, M.; Kratochvil, P. *Macromolecules* **1996**, *29*, 6023 - 6030.

Fluorescence and UV Absorption Spectroscopy

Chapter 6

Application of Size-Exclusion Chromatography with Fluorescence Detection to the Study of Polymer Reaction Kinetics

Ashok J. Maliakal[1], Ben O'Shaughnessy[2], and Nicholas J. Turro[1,*]

Departments of [1]Chemistry and [2]Chemical Engineering,
Columbia University, New York, NY 10027

SEC with fluorescence detection has been used to study the kinetics of polymer chain end reactions in solution. The presence of a fluorescent pyrene label in a styrene terminated polystyrene allows for sensitive detection of both starting material and product, which are resolved as a function of molecular weight through SEC. Hence measurements can be performed under pseudo-first order conditions allowing a simple determination of bimolecular polymer end-end reaction rate constants. This method is applicable to measurement of both activation controlled and diffusion controlled polymer reaction kinetics (i.e. radical-radical reactions).

The measurement of polymer end-end reaction rates has been of interest to experimental(1,2) and theoretical polymer scientists.(3-5) Fundamental theories have been developed to predict the effect of chain length on reaction rates.(4,6) However, several experimental difficulties have hindered the measurement of polymer chain end reaction kinetics (e.g., synthesis of suitable monodisperse end-labeled samples and low sensitivity of analytical techniques for the detection of polymer end groups). Although photophysical approaches have been used to measure interpolymer chain end reaction rates,(2,7-10), these methods are limited in dynamic range by the fluorescence and phosphorescence lifetimes of the probes. In order to address the issues of sensitivity and dynamic range, an SEC-based method has been developed which resolves reactant and product based on molecular weight and which employs fluorescence detection in the analysis of the disappearance of starting material and the formation of the end to end coupling product.(11,12) Previous studies of the kinetics of activation controlled polymer end-end reactions observed either loss of starting material,(13) or product formation,(14) but were unable to simultaneously measure both.

The current chapter discusses the application of SEC with fluorescence detection (SEC-F) to the study of both activation and diffusion controlled polymer chain end reaction kinetics. The chain length dependence has been determined for the activation controlled reaction of polystyryl lithium **1** [PSLi] (see Figure 1; N = degree of polymerization) with a second polystyrene molecule **2** (referred to as PSPYSTY) which is selectively labeled with one pyrene fluorophore and contains a reactive styrene functionality at the chain end.(11)

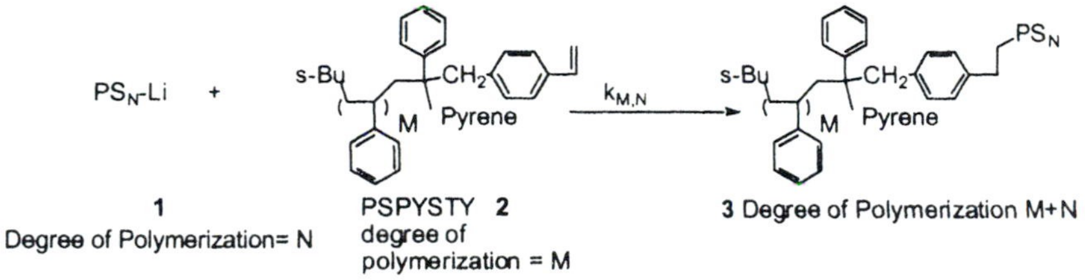

PS_N-Li   +

1
Degree of Polymerization= N

PSPYSTY **2**
degree of
polymerization = M

$k_{M,N}$

3 Degree of Polymerization M+N

Figure 1. Synthesis of Styrene Endlabeled-Pyrene-Labeled Polystyrene, PSPYSTY (2).

The high sensitivity of the pyrene probe **2** permits measurement of pseudo-first order rate constants for polymer end-end reactions, allowing for relatively simple kinetic analysis. Surprisingly this study reveals an increase in the bimolecular reaction rate constant with increasing molecular weight. This result has been discussed in the context of the complex supramolecular environment present in the PSLi/benzene system.(11)

In switching to a fast reacting system (approaching diffusion controlled), mixing of reagents to start the reaction, and time resolution become important issues. Time-resolved detection methods exist (i.e., time resolved IR,(15) time resolved ESR, (16) and laser flash photolysis followed by time–resolved UV-Vis spectroscopy(17) and photophysical quenching techniques.)(7-9,18-20)) However these techniques again involve significant signal to noise issues and are limited by the excited state lifetime of the probe employed. In the case of photophysical methods, relating quenching rate constants to reaction rate constants is complicated by the fact that quenching rate constants occur at a distance due to dipole-dipole mechanisms,*(17)* where as chemical reaction requires close contact between chain ends to accomplish reaction.*(1)* To address these issues, photoinitiators have been used to generate fluorescent labeled polymers containing a chain end radical (macroradicals), and the reactions of these macroradicals have been monitored using SEC-F. The dynamic range in this case is no longer an issue, since fluorescence is used only as a label whose transfer from starting material to product is being monitored after completion of reaction. To address the issue of time resolution, a method of competitive kinetics is proposed for extracting the rate dependence from these fast polymer chain end reactions. Although actual rate constants and chain length dependences have not yet been achieved, the viability of the SEC-F method in the case of fast reactions is demonstrated.

Experimental

Materials. Unless specified, compounds were purchased from Aldrich. Styrene was dried over calcium hydride, distilled onto dibutylmagnesium, and then distilled under reduced pressure fresh prior to use. Benzene was dried over calcium hydride, distilled over PSLi, and distilled fresh prior to use. THF (Acros) was dried over potassium hydride. 1-(1-Phenyl-vinyl)-pyrene (**4**) was synthesized using a modified literature procedure.(21). Synthesis of Tosylate **6** is described elsewhere.*(22)* The synthesis of PSPYSTY (number averaged molecular weight M_N =2300, degree of polymerization M = 22) has been described previously.*(11)*

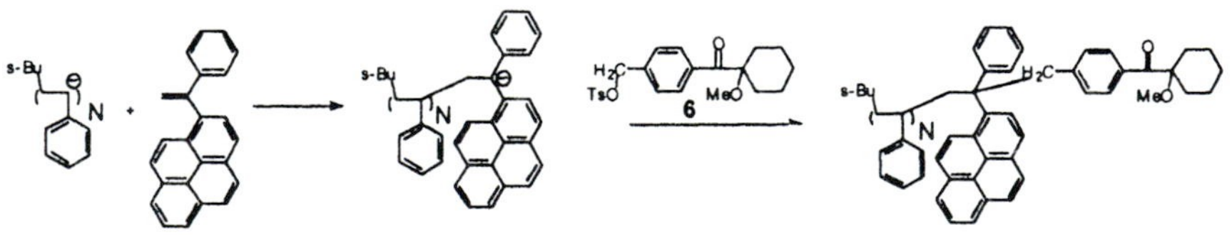

Figure 2. Synthesis of PSPYPI 4.

Synthesis of PSPYPI (7). 1-(1-Phenyl-vinyl)-pyrene **4** was freeze-dried from dry degassed benzene under vacuum. After this 10 ml of dry degassed THF is added. In a separate flask, the tosylate **6** was freeze-dried from dry degassed benzene as well; 2 mL of dry degassed THF is added to this flask. PSLi was synthesized by addition of 0.65 mmol of s-butyl lithium to 1.9 mL of styrene (16.25 mmol) in 5 mL of benzene which is cooled in an ice bath. Styrene and benzene were both dried over calcium hydride and freshly vacuum distilled prior to use. The appropriate amounts of each reagent are transferred via syringe into the flame dried reaction flask. The reaction is allowed to warm to room temperature an aged for 2 hours (reddish orange solution due to PSLi anion). 2 mL of the resulting orange solution is transferred via flame-dried syringe to the flask containing a THF solution of **4**. A dark blueish purple solution is obtained instantly, this solution is aged 10 mins and then transferred via cannula to the solution of **8** in THF held at −78°C using a dry ice acetone bath. The blueish-purple solution (presumable the di-aryl-anion **5**) decolorizes rapidly. The reaction is monitored by TLC, and **6** is observed to disappear. PSPYPI is recovered as a pale yellow solid after several precipitations from methanol. SEC indicates one peak with polydispersity 1.2, $M_N = 2000$. UV-Vis and fluorescence spectra confirm incorporation of pyrene into polymer. UV-Vis and excitation spectra do indicate the presence of a highly colored impurity. ^{1}H-NMR is consistent with the product. The benzylic proton resonances are split from the expected singlet due to diastereotopic interactions from the atactic polystyrene backbone known to extend over up to 3 styrene residues. *(23)(24)*

Instrumentation. Size exclusion chromatography was performed on a Polymer Labs SEC with a GTI/Spectrovision FD-500 fluorescence detector. 2 PL mixgel 5_ C and 1 PL gel 5_ 100A columns were used in series, and calibration was performed with polystyrene standards (Polymer Labs).

Measurement of the Rate of Reaction of 1 with 4. Polystyryl lithium **1** is generated by reaction of an appropriate amount of s-butyllithium with 5.85 mL (50 mmol) of styrene in 10 mL of dry degassed benzene. SEC molecular weight (M_N) of resultant PSLi is used to confirm the concentration of s-butyl lithium added. After reaction is complete (~2 hrs), this solution is maintained at constant temperature of $30 \pm 0.2°C$ using an IKA ETS-D4 temperature controller. In a separate flask, PSPYSTY (**2**) (5 mg, 2.2×10^{-3} mmol) is dissolved in a small amount of dry degassed benzene, and added to the solution of **1** in benzene. The addition time is marked with a stopwatch, and aliquots are removed subsequently and quenched onto methanol. A small degree of coupling reaction (few percent) is observed due to the formation of free radical chains ends resulting from oxidation of anionic chains, since the methanol employed is not rigorously degassed (see shoulder in Figure 3 for product peak **3**).*(25)* However, this is not expected to affect the kinetic analysis. The samples thus generated are dissolved in THF and filtered through a short plug of silica gel to remove lithium salts prior to injection into the SEC.

Reaction of PSPYSTY (2) and (diphenyl-phosphinoyl)-(2,4,6-trimethyl-phenyl)-methanone (15) in Rayonet Reactor

PSPYSTY (**2**) was dissolved in benzene (spectroscopic grade Aldrich). The O.D. for solution is 1.78 @313 nm, which translates to 1.2×10^{-4} M PSPYSTY based on pyrene absorption ($\varepsilon_{py} = 1.5 \times 10^{4}$ M in nonpolar solvent)(26) **8** was added (O.D. 0.07 at 381 nm) and mixture is degassed by argon bubbling, and irradiated for 3 mins at 420 nm in Rayonet Carousel Photoreactor with 420 nm lamps (Strontium Pyrophosphate/Europium lamps emission max at 420 nm, 34 nm width at 50% height. manufactured by Southern New England Ultraviolet, Branford CT)

UV-VIS absorption indicates complete consumption of **8** and yellowing of the solution is observed. Reaction is continued by adding more **8** (0.07 Abs at 381 nm). After further irradiation (~3 mins), reaction is stopped and products injected onto SEC (see Figure 11). SEC fluorescence conditions (excite at 330 nm, emission at 450 nm).

Results and Discussion

Measurement of Activation Controlled Polymer Chain End Reaction Rates

Previous efforts*(11)* from our research group have measured the rate of polymer chain end reactions as a function of chain length using SEC-F for the activation controlled   reaction of PSLi (degree of polymerization N) with a macromolecular fluorescent labeled styrene (PSPYSTY; **2**). Reaction of **1** with fluorescently labeled PSPYSTY (**2**) is clean and yields only one product **3** after quenching with methanol (see Figure 1). Both PSPYSTY and **3** can be resolved by SEC as a function of molecular weight, and observed by fluorescence detection (excitation at 330 nm, emission at 400 nm) in a regime in which there is negligible background signal from polystyrene.

The reaction illustrated in Figure 1 was employed for measurement of rate constants for the end to end polymer coupling.*(11)* PSLi solutions of varying chain lengths were heated to 30°C and kinetics were measured at this temperature. A solution of PSPYSTY (**2**) dissolved in a small amount of benzene was added to the PSLi solution. Aliquots were withdrawn periodically and quenched into methanol. The resulting samples were injected into the SEC to produce a time series of SEC traces (representative examples are shown in Figure 3). The fluorescence of the pyrene is observed in the starting material PSPYSTY (**2**) and product **3**, but the PSLi is effectively non-emissive under these detection conditions (excitation at 330 nm and observation at 400 nm). For this reason we are able to operate under pseudo-first order kinetic conditions with a 50-100 fold excess of PSLi as compared to PSPYSTY (**2**) without any significant interference in the SEC fluorescence analysis from the PSLi. The area integrations of peaks **2** and **3** in Figure 3 are proportional to the concentrations of starting material and product in the reaction. These areas are used to calculate the fraction conversion as a function of time (see Figure 4). The conversion data for the starting material and product are fit with exponential curves (see Figure 4).

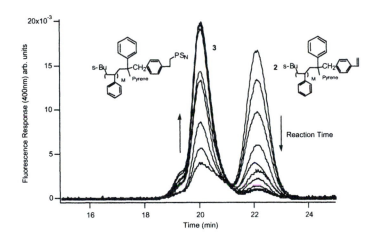

Figure 3. SEC Traces with Fluorescence Detection (Excitation at 330 nm, Observation at 400 nm) at Several Reaction Times. Peak for Starting Material 2 Decreases with Reaction Time and Peak for Products 3 Increases with Reaction Time. Reaction Conditions: Benzene, 30°C. (Adapted from Reference 11.) (Copyright 2003 American Chemical Society.)

As the chain length was varied, the concentration of PSLi also changed, and for this reason it was important to assess the dependence of the rate constant on PSLi concentration. In order to confirm the PSLi concentration dependence (which has been reported of the order of 0.5 in polystyrene propagation*(27-29)* as well as varying between 0.48 to 0.87*(30)*), the dependence of k_{obs} on [PSLi] at a constant molecular weight ($M_N = 10$ K; $N \approx 100$) was measured. Although there is much discussion in the literature regarding aggregation state and reaction mechanism, under our experimental conditions we observe a roughly linear dependence of k_{obs} *versus* [PSLi] which is consistent with the rate equation 1.*(11)*

$$k_{obs} = k_{M,N}[PSLi]^{0.9 \pm 0.1} \qquad (1)$$

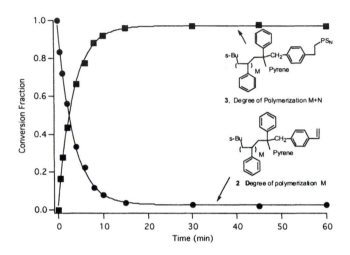

*Figure 4. Representative Kinetic Trace and First Order Exponential Fit for Decrease of Starting Material **2** and Growth of Product **3** as Measured from Area of Peaks in Figure 1. (M_N for PS_NLi (**1**) = 4500 ± 100 amu, M_N for **3** (M + N) = 6900 ± 200) (Adapted from Reference 11.) (Copyright 2003 American Chemical Society.)*

This experimentally determined relationship was used to calculate the bimolecular rate constant of reaction $k_{M,N}$ as a function of chain length at constant volume fraction polystyrene. (In order to simplify units, the approximation of a first order dependence in [PSLi] was made in calculating $k_{M,N}$ from k_{obs}). These results are presented in Figure 5. It is interesting to note that the bimolecular rate constant $k_{M,N}$ increases as the degree of polymerization increases from N = 31 to 246 (M_N = 3.3 K-25.6 K) from 3.6 $M^{-1}min^{-1}$ to 10.3 $M^{-1}min^{-1}$ ± 1 $M^{-1}min^{-1}$. The dashed line indicates the transition between dilute and semi-dilute which occurs for polystyrene at this concentration in benzene at approximately 4.6 K.

Previous studies have employed the use of SEC-F to explore coupling processes in high temperature melt processing *(31-33)* and at the thin film interface.*(12)* Our work has extended the SEC-F detection methodology to measure bimolecular rate constants directly for activation controlled polymer-polymer reactions in solution.

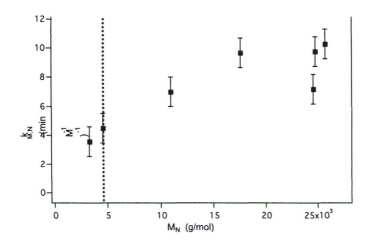

Figure 5. Chain Length Dependence of Bimolecular Polymer End-End Reaction Rate Constant $k_{M,N}$ (Plotted with Estimated Error ± 1 $M^{-1}min^{-1}$). M_N of PSPYSTY 2 = 2300 ± 100, M = 22. M_N of PSLi (1) Varied from 4K-25K (N = 40-240). Dotted Line Indicates Transition Between Dilute and Semi-dilute Regimes (Adapted from Reference 11.) (Copyright 2003 American Chemical Society.)

Several possible reasons were considered for the increase of $k_{M,N}$ with chain length.*(11)* However, a change in the supramolecular structure of **1** with increasing chain length was most consistent with the experimental data.*(30,34-36)* In this hypothesis, as N increases, the aggregation number of supramolecular micellar aggregates of **1** are expected to decrease due to increased excluded volume repulsive interactions of the polystyryl chains.*(35,36)* There is a possibility that in these smaller micelles, **1** is more reactive (See Figure 6).*(36)*

Attempts to investigate kinetics at longer chain lengths were limited by an increase of viscosity to the point where reaction was limited by stirring. For example, an experiment performed with M_N = 55K PSLi chains (30% by volume PSLi) resulted in a viscous solution where stirring was no longer efficient. PSPYSTY (**2**) did not mix homogeneously as evidenced by the observation of nonexponential and erratic kinetics for disappearance of starting material and appearance of product. At this chain length and concentration, the PSLi solution

122

behaves like an entangled transient network, as predicted by polymer theory.(3) The stir rate (~100-800 min⁻¹) is now faster than the relaxation of the polymer network, so in order for stirring to occur, the stir bar must either break the polymer chains or wait for entangled chains to relax. The dependence of fractional conversion on the location of sampling indicates incomplete mixing. Transient network formation presents a significant challenge to the study of polymer reactions in the entanglement region, a problem similar to the study of polymer-polymer reactions in the solid state.(2,37)

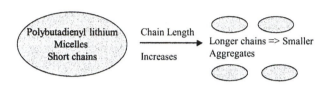

Figure 6. Effect of Chain Length on Polybutadienyl Lithium Micellar Aggregation State.

Despite the high activation energy for addition of PSLi to the styrene moiety, the reaction becomes diffusion controlled due to the slow polymer chain end diffusion under entanglement constraints. Approaches to studying diffusion controlled polymer reactions using SEC-F are discussed in the next section.

Approaches to Measuring Diffusion Controlled Polymer Reaction Kinetics

Fast reaction processes can be measured through competitive kinetics. Concurrent reactions of mixtures have been used to assess the relative rate constants between a reactive intermediate and two or more different competitive reaction partners.(38) The system envisioned for measuring the chain length effect for high Q rate constants is based on this concept of the kinetics of concurrent reactions of mixtures. A monodisperse fluorescent labeled photoinitiator end-labeled polymer (P-R) (degree of polymerization M) was sought which upon photolysis yields a macroradical P· (see Figure 7). This macroradical is formed in the presence of a variable but high concentration of a reactive small molecule (S) as well as a variable but high concentration of an analogous polymeric reactive molecule (M) (degree of polymerization N). The simplified kinetic scheme is illustrated in Figure 7.

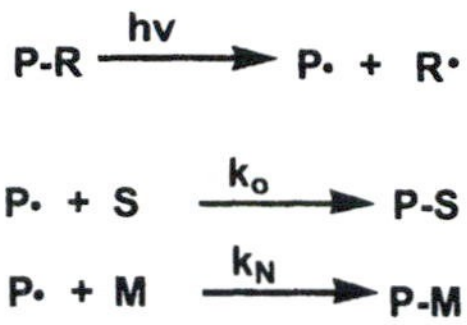

$$P\text{-}R \xrightarrow{\ h\nu\ } P\cdot + R\cdot$$

$$P\cdot + S \xrightarrow{\ k_o\ } P\text{-}S$$

$$P\cdot + M \xrightarrow{\ k_N\ } P\text{-}M$$

Figure 7. Concurrent Reaction of P· with competitive small (S) and macromolecular (M) reagents.

After a reaction time t, measurement of the relative amounts of *P-M* (degree of polymerization M+N) and *P-S* (degree of polymerization M) would be performed using SEC with fluorescence analysis. The ratio of *P-M* to *P-S* could be related to the ratio of rate constants k_N/k_o through equation 2. The reaction of *P·* with itself or with *R·* could be minimized through high relative concentrations of S and M (similar to the pseudo-first order kinetic conditions employed in the previous section).

$$\frac{k_N}{k_o} = \frac{[PM][S]}{[PS][M]} \tag{2}$$

Performing the experiment at various degrees of polymerization for *P-R* and M would permit assessment of the chain length dependence of this fast reaction.

With this kinetic model in mind, two approaches were attempted to generate a suitable macroradical *P·*. In the first approach, the monodisperse photoinitiator terminated pyrene labeled polystyrene PSPYPI (7) was synthesized. The commonly used α-alkoxy-ketone photoinitiator moiety was incorporated into the polymer 7 (see Figure 8).*(39,40)(41)* Typically this photoinitiator, upon absorption of light, undergoes rapid intersystem crossing and Norrish Type I cleavage to form the macroradical-radical pair illustrated in Figure 8. However triplet energy transfer from the photoinitiator to the pyrene moiety prevented PSPYPI from cleaving, rendering the photoinitiator inactive.*(22)* If the photoinitiator is separated from the chromophore, then it should be photoactive in the presence of a pyrene chromophore, as was found to be the case using an external photoinitiator.

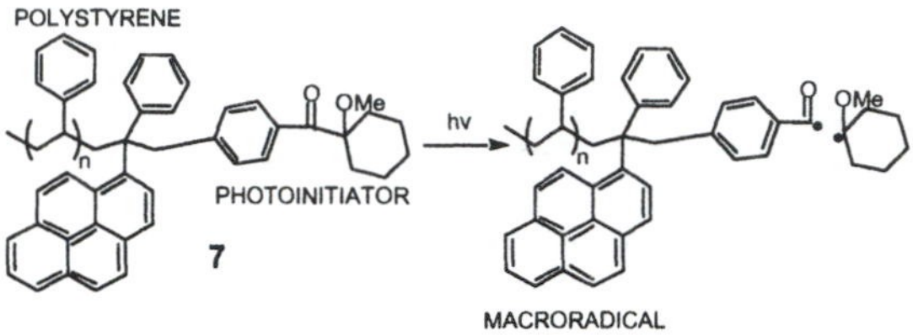

Figure 8. Compound chosen as P-R for competitive kinetics to measure k_N/k_o.

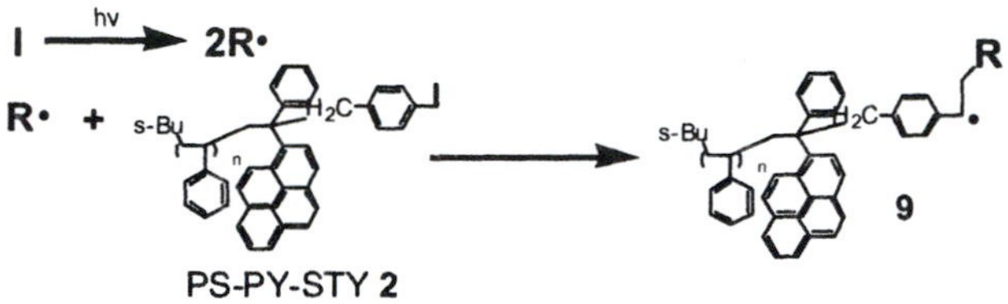

Figure 9. External Photoinitiation to generate macroradical PSPYPI.

In a second strategy, the macroradical was generated through reaction of an external photoinitiator (I, see Figure 9) with PSPYSTY 2. In this strategy an external photoinitiator was photolysed independent of the polymeric species. The resulting radicals add rapidly to the styrene terminated polystyrene (PSPYSTY) to generate the macroradical illustrated in Figure 9. The fate of the macroradical 9 can be tracked with the aid of the fluorescent label using SEC analysis.

There are several different types of radical photoinitiators.*(42)* The following considerations were of prime importance in our selection of an external photoinitiator. First, a long wavelength absorption was desirable which could be selectively excited independent of PSPYSTY. Secondly, we sought a radical which adds rapidly to styrene in order to rapidly generate the macroradical 9. These two criteria coupled with the desire for a photoinitiator with a high quantum yield lead us to choose (diphenyl-phosphinoyl)-(2,4,6-trimethyl-phenyl)-methanone (8) as the external photoinitiator.

Excitement of 8 can be achieved independent of PSPYSTY due to the long wavelength absorption band of 8 (at 420 nm). Furthermore literature references*(42)* indicate that the resultant phosphinoyl radicals (see Figure 10) add to the styrene moiety at a rate of $\sim 10^7$ $M^{-1}s^{-1}$. Thus external initiation is another route to generating monodisperse macroradicals rapidly. The resultant macroradicals (**10**) are capable of dimerizing (or undergoing disproportionation) as may be observed from SEC analysis of reaction (see Figure 11).

After exhaustive photolysis in THF, an aliquot of the reaction mixture is diluted and injected into the SEC. The early peak in Figure 11 (retention time 20.7 minutes) corresponds to a molecular weight ($M_N = 5000 \pm 100$) consistent with the dimeric product **11** in Figure 10. The latter peak (retention time 21.9 minutes) corresponds to the molecular weight of the starting material **2** (2300 ± 100 amu). This peak most likely corresponds to the product resulting from the disproportionation reaction of the macroradical **10** and/or reaction of **10** with 2,4,6-trimethylbenzoyl radical (see Figure 10).

Although the dimerization reaction of **10** shows in concept the potential of using fluorescence SEC to monitor kinetics of polymer chain end reactions in the fast reaction regime, these particular reaction conditions involve too many competing reactions to allow for the simple extraction of k_N the chain length dependent interpolymer end reaction rate constant.

Our lab is currently seeking suitable compounds for the roles of S and M in Figure 7. The nitroxides which have been employed previously as radical traps are promising candidates (see Figure 12).*(43,44)*

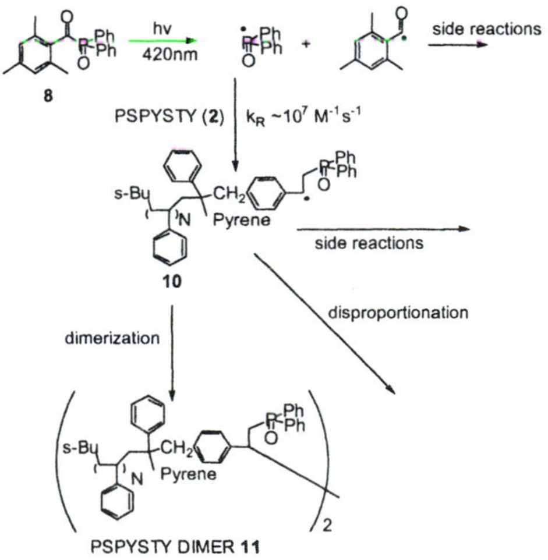

Figure 10. Reaction Pathway for Photolysis of 8 in the presence of PSPYSTY (2).

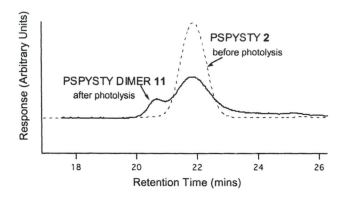

*Figure 11. SEC Analysis of Photolysis of **8** in the Presence of PSPYSTY **2** (see Figure 10 for proposed reaction pathways).*

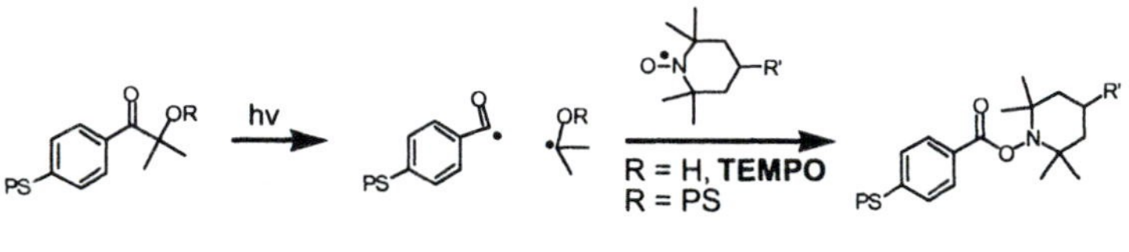

Figure 12. Plan for Using Nitroxide Based Radical Traps (R' = H (TEMPO), or polystyrene, R' = PS) to Trap Macroradical.

Summary and Conclusions

SEC based measurement of kinetics using fluorescence detection is demonstrated to be a powerful method to extract bimolecular rate constants of polymer end-end reactions. The sensitivity of fluorescence detection permits kinetic study under pseudo-first order conditions, allowing for a simple kinetic analysis. This method has been applied to the problem of measuring the chain length dependence of polymer end-end reaction rates.

Attempts to measure kinetics in the entanglement regime were thwarted by the inability to mix starting material and product efficiently to create homogenous reaction. Nonexponential kinetics result due to heterogeneity in reaction environment, a situation similar to that observed for polymer reactions in the solid state.*(2,37)*

For fast reactions, photoinitiation gets around the problem of mixing dependent reaction rates by premixing reagents and homogeneously initiating the reaction using light. A concurrent competing reaction scheme was proposed which would allow for determination of the chain length dependence of inter polymer chain end kinetics.

An initial approach to *P-R* involving the attachment of a photoinitiator in close proximity to a pyrene end-labeled polymer failed due to quenching of the photoinitiator by the pyrene. Subsequent attempts employing an external photoinitiator allowed for the formation of fluorescent labeled macroradical as observed by SEC-F. Although the generation and subsequent reactions of PS-macroradicals **11** could potentially be studied kinetically, the requirement of a large amount of external photinitiator **8** complicated the kinetic scheme.

The current work demonstrates that fast reactions can be monitored by SEC-F. In order to make these systems more transparent to kinetic analysis, we are currently altering the distance between photinitiator and fluorescent label in order to minimize triplet energy transfer quenching in the photoinitiator. A second possible solution to get around utilizing external photoinitiators is to use a fluorophore with a triplet energy higher than the photoinitiator. This would make triplet energy transfer thermodynamically unfavorable and permit cleavage of the photoinitiator triplet. Furthermore efforts are underway to synthesize suitable macromolecular radical traps (*M*) in order to be able to measure fast interpolymer chain end reaction rates using the concurrent reaction scheme illustrated in Figure 7.

Acknowledgements

The authors thank the National Science Foundation (Grants CHE-00-91460 and CHE-01-10655) for its support of this research. AJM thanks the NSF for a graduate fellowship.

References

(1) de Kock, J. B. L.; Van Herk, A. M.; German, A. L. *J. Macromol. Sci.-Polym. Rev* 2001, *C41*, 199-252.

(2) Mita, I.; Horie, K. *J. Macromol. Sci., Rev. Macromol. Chem. Phys.* 1987, *C27*, 91-169.

128

(3) de-Gennes, P. *Scaling Concepts in Polymer Physics*; Cornell University Press: Ithaca, NY, 1985.

(4) Friedman, B.; O'Shaughnessy, B. *Int. J. Mod. Phys. B.* 1994, *8*, 2555-2591.

(5) Khokhlov, A. *Makromol. Chem. Rapid. Commun.* 1981, *2*, 633.

(6) Wang, Y. C.; Morawetz, H. *Macromolecules* 1990, *23*, 1753-1760.

(7) Winnik, M.; Sinclair, A.; Beinert, G. *Macromolecules* 1985, *18*, 1517-1518.

(8) Sinclair, A.; Winnik, M.; Beinert, G. *J. Am. Chem. Soc.* 1985, *107*, 5798-5800.

(9) Gebert, M.; Torkelson, J. *Polymer* 1990, *31*, 2402-2410.

(10) Yu, D.; Torkelson, J. *Macromolecules* 1988, *21*, 852-853.

(11) Maliakal, A.; Greenaway, H.; O'Shaughnessy, B.; Turro, N. J. *Macromolecules* 2003, *36*, 6075 - 6080.

(12) Moon, B.; Hoye, T.; Macosko, C. *J. Polym. Sci. Part. A. Polymer Chemistry* 2000, *38*, 2177-2185.

(13) Okamoto, A.; Shimanuki, Y.; Mita, I. *Eur. Polym. J.* 1982, *18*, 545-548.

(14) Okamoto, A.; Toyoshima, K.; Mita, I. *Eur. Polym. J.* 1983, *19*, 341-346.

(15) Sluggett, G.; Turro, C.; George, M.; Koptyug, I.; Turro, N. *J. Am. Chem. Soc.* 1995, *117*, 5148-5153.

(16) Liu, Z.; Weber, M.; Turro, N.J.; O'Shaughnessy, B. in "Photoinitiated Polymerization", eds. K.D. Belfield and J.V. Crivello, ACS Symposium Series 847, Washington, DC 2003.

(17) Turro, N. J. *Modern Molecular Photochemistry*; University Science Books: Sausalito, 1991.

(18) Horie, K.; Mita, I. *Macromolecules* 1978, *11*, 1175-1179.

(19) Mita, I.; Horie, K.; Takeda, M. *Macromolecules* 1981, *14*, 1428-1433.

(20) Gebert, M. S.; Yu, D. H. S.; Torkelson, J. M. *Macromolecules* 1992, *25*, 4160-4166.

(21) Quirk, R.; Schock, L. *Macromolecules* 1991, *24*, 1237-1241.

(22) Maliakal, A. *Ph. D. Thesis, Dept. of Chemistry;* Columbia University: NY, 2003.

(23) Moad, G. *Chem. Aust.* 1991, *58*, 122-126.

(24) Bevington, J.; Lyons, R.; Senogles, E. *Eur, Polym. J.* 1992, *28*, 283-286.

(25) Fetters, L.; Firer, E. *Polymer* 1977, *18*, 306-307.

(26) Murov, S.; Carmichael, I.; Hug., G. *Handbook of Photochemistry*; M. Dekker: New York, 1993.

(27) Worsfold, D.; Bywater, S. *Can. J. Chem.* 1960, *38*, 1891-1900.

(28) Morton, M.; Fetters, L. *J. Polym. Sci. Part. A. Polymer Chemistry* 1964, *2*, 3311-3326.

(29) Morton, M.; Fetters, L.; Pett, R.; Meier, J. *Macromolecules* 1970, *3*, 327-332.

(30) Fetters, L. J.; Huang, J. S.; Stellbrink, J.; Willner, L.; Richter, D. *Macromol. Symp.* 1997, *121*, 1-26.

(31) Gray, M.; Kinsinger, M.; Torkelson, J. *Macromolecules* 2002, *35*, 8261-8264.

(32) Yin, Z.; Koulic, C.; Jeon, H.; Pagnoulle, C.; Macosko, C.; Jerome, R. *Macromolecules* 2002, *35*, 8917-8919.

(33) Schulze, J.; Moon, B.; Lodge, T.; Macosko, C. *Macromolecules* 2001, *34*, 200-205.

(34) Stellbrink, J.; Willner, L.; Jucknischke, O.; Richter, D.; Lindner, P.; Fetters, L. J.; Huang, J. S. *Macromolecules* 1998, *31*, 4189-4197.

(35) Stellbrink, J.; Willner, L.; Richter, D.; Lindner, P.; Fetters, L. J.; Huang, J. S. *Macromolecules* 1999, *32*, 5321-5329.

(36) Stellbrink, J.; Allgaier, J.; Willner, L.; Richter, D.; Slawecki, T.; Fetters, L. J. *Polymer* 2002, *43*, 7101-7109.

(37) Horie, K.; Mita, I. *Adv. Polym. Sci.* 1989, *88*, 77-128.

(38) Espenson, J. *Chemical Kinetic and Reaction Mechanisms*; McGraw Hill: New York, 1981.

(39) Fouassier, J. *Euro. Coatings Journal* 1996, 723-726.

(40) Turro, N.; Wu, C. *J. Am. Chem. Soc.* 1995, *117*, 11031-11032.

(41) Wu, C. *Ph. D. Thesis, Dept. of Chemistry;* Columbia University: New York City, 1994.

(42) Reetz, I.; Yagci, Y.; Mishra, M. *Handbook of Radical Vinyl Polymerization*; Marcel Dekker, Inc.: New York, 1998; Vol. 48.

(43) Braslau, R.; Anderson, M.; Rivera, F.; Jimenez, A.; Haddad, T.; Axon, J. *Tetrahedron* 2002, *58*, 5513-5523.

(44) Moad, G.; Solomon, D. *Chemistry of Free Radical Polymerization*; Pergamon: Oxford, U.K., 1995.

Chapter 7

Characterizing Glycoproteins and Protein–Polymer Conjugates with Light Scattering, UV Absorbance, and Differential Refractometry

Brent S. Kendrick

Amgen, Inc., MSAC–24F, 4000 Nelson Road, Longmont, CO 80503

Recent developments in size exclusion high performance liquid chromatography with on-line static light scattering (LS), refractive index (RI), and ultraviolet (UV) detection allow rapid solution-state characterization of glycoproteins and protein-polymer conjugates. Through simple mathematical relationships of peak areas from LS, RI, and UV detectors, the molecular mass, degree of glycosylation or polymer conjugation, and the solution association state can be readily determined. Baseline resolution of the chromatographic peaks is not required; peaks need only be sufficiently separated to represent relatively pure fractions.

Characterizing the solution-state molecular mass and association state of conjugated proteins such as glycosylated and polyethylene glycol (PEG) modified proteins is important during development of recombinant therapeutic protein drugs. Determining the degree of glycosylation can be used during initial development of a glycoprotein, optimization of fermentation and purification processes to ensure the desired level of glycosylation is achieved. Similarly, determining the degree of PEGylation is important to ensure the conjugation chemistry and purification techniques lead to the desired product. Simultaneously characterizing the extent of glycosylation/PEGylation in solution and association state leads to an understanding of the aggregation behavior of the molecule, the reversibility of such associations in solution, and can be used to track the stability of the molecule. Empirical methods such as size exclusion chromatography, sedimentation velocity or dynamic light scattering can be run under various solution conditions, but only measure hydrodynamic radius. Sedimentation equilibrium (SE) ultracentrifugation is another useful technique, however it relies on analysis of extremely pure samples that are stable for the length of the analysis, which can take up to several days *(1, 2)*.

There are some excellent reports on using size exclusion high performance liquid chromatography (SEC-HPLC) coupled on-line with a combination of refractive index (RI) detection and ultraviolet (UV) detection for conjugated protein characterization *(3-6)* and protein association states *(7, 8)*. Determining degree of conjugation and association state by these methods is a lengthy process (for a review see: refs. *(7, 8)*). Briefly, a selected number of proteins are used as mass standards to generate a calibration curve of (LS) (UV) $(RI)^{-2} \varepsilon_p^{-1}$ vs. molecular mass (where ε_p is the polypeptide extinction coefficient). The curve is then used to derive the association state (degree of oligomerization) of the conjugate either with itself or with protein receptors / ligands (LS is the area under the LS peak). Through a self-consistent iterative process the degree of conjugation can also be determined.

This chapter covers recent improvements *(9)* using simple mathematical relationships to facilitate rapid solution-state characterization of glycoproteins and protein-polymer conjugates. Time consuming iterative procedures have been replaced with a simple relationship using an arbitrary molecular mass value determined by light scattering combined with the corresponding arbitrary (and unknown) conjugated protein *dn/dc*. This relationship is discussed in detail below and applied in various situations to determine degree of conjugation and association state of protein-conjugates.

Instrument setup and detector calibration

The following is a brief description of the method *(9)* we use on a Wyatt MiniDAWN with ASTRA software (Wyatt, Inc., Santa Barbara, CA) laser light scattering instrument. Analog inputs are obtained from an Agilent 1100 HPLC system with UV (variable wavelength) and RI detectors. The solvent flows through the detectors in the following order: UV, LS, RI.

- Inject 50 µg bovine serum albumin (BSA) over a Tosohaas G3000SWXL column in a mobile phase of 10 mM sodium phosphate, 140 mM sodium chloride, pH 7.0 (PBS) at a flow rate of 0.5 ml/min, collecting RI, UV (280nm), and LS signals.

- Determine UV and RI detector calibration constants. *UV cal const$_A$* = (0.1 *A*/volt)/(1000 mg/g), and *A* is absorbance units. Integrate the monomer peak area for each of the RI, UV, and LS signals. For example, with the Wyatt system, the signals are given in volts, and peak areas in volts*min. Using the known *(8)* *dn/dc* (i.e. the differential change in solution refractive index as a function of concentration, *c* (g/ml)) for BSA in PBS of 0.187 ml/g, and the extinction coefficient at 280 nm, ε_{BSA} = 0.670 ml/mg/cm, the following equality applies:

$$(dn/dc)_{BSA} = \frac{(RI\ area)(RI\ cal\ const)}{(UV\ area)(UV\ cal\ const_A)} \varepsilon_{BSA} \quad (1)$$

which is rearranged to give the RI calibration constant:

$$RI\ cal\ const = \frac{(dn/dc)_{BSA}(UV\ area)(UV\ cal\ const_A)}{\varepsilon_{BSA}(RI\ area)} \quad (2)$$

- Unknown protein *dn/dc* values may then be determined from the calibrated detectors and the following form of Eq. 3:

$$(dn/dc)_p = \frac{(RI\ area)(RI\ cal\ const)}{(UV\ area)(UV\ cal\ const_A)} \varepsilon_p = (dn/dA)\varepsilon_p \quad (3)$$

Care must be taken to minimize band-broadening effects by being constistent in peak area selection *(9)*.

- The calibration constant for the LS detector can be determined using the following relationship and the LS vendor's software:

$$LS\ cal\ const_{true} = LS\ cal\ const_{arb} \frac{true\ molecular\ mass\ of\ BSA}{arbitrary\ molecular\ mass\ of\ BSA} \quad (4)$$

where the true molecular mass of BSA monomer is 66,270 Da *(8)*. For molecular weight calculations in ASTRA the UV instrument calibration constant must be converted to the correct concentration units: *UV cal const$_p$* = *UV cal const$_A$(dn/dc)$_p$/ε_p*.

Theoretical

The refractive index (n) of a substance is a function of its chemical groups, and independent of higher order structures. Thus, the dn/dc for a protein-conjugate species $(dn/dc)_{cp}$ is $(4, 7)$:

$$\left(\frac{dn}{dc}\right)_{cp} = \frac{M_p}{M_{cp}}\left(\frac{dn}{dc}\right)_p + \frac{M_c}{M_{cp}}\left(\frac{dn}{dc}\right)_c \tag{5}$$

where M is molecular mass (Da), n is the refractive index, and the subscripts p, c, and cp refer to the protein, conjugating species, and conjugated protein respectively. Using the conservation of mass equation: $M_c = M_{cp} - M_p$, and substituting into Eq. 5 gives the following equation upon rearrangement:

$$M_{cp} = M_p \frac{\left(\dfrac{dn}{dc}\right)_p - \left(\dfrac{dn}{dc}\right)_c}{\left(\dfrac{dn}{dc}\right)_{cp} - \left(\dfrac{dn}{dc}\right)_c} \tag{6}$$

$(dn/dc)_p$ and $(dn/dc)_c$ can be determined from independent experiments on the protein before conjugation and on the pure conjugating species. The degree of association (N) is given by:

$$NM_{mono,\,cp} = M_{cp} \text{ and } NM_{mono,\,p} = M_p \tag{7}$$

Where $M_{mono,\,p}$ is typically determined by the known sequence.

Determining association state of a conjugated protein

The simplified classical light scattering equation has the form $(Kc/R \approx 1/M)$, where K is a constant proportional to $(dn/dc)^2$, and R is the excess intensity of light scattered by the solution over that of the solvent (10). The key finding that led to the simplification of determining conjugated protein association states was the observation that the product $M \bullet (dn/dc)$ is a constant (α) (dependent on a given protein) when M is calculated by the analysis software and the RI signal is used as the concentration detector: $c = (n-n_0)/(dn/dc)_{cp}$ where n is the RI signal for a given position in the eluting peak, and n_0 is the RI signal due to solvent alone. Since it the product we are after, any dn/dc value, $(dn/dc)_{arb}$, can be used during light scattering calculations to give an arbitrary M_{cp} for a selected chromatographic peak. This product can be written as:

$$M_{cp,arb} * (dn/dc)_{cp,arb} = M_{cp,true} * (dn/dc)_{cp,true} = \alpha_{cp} \tag{8}$$

Which can be rearranged to yield the true *polypeptide* portion of the molecular mass by LS, RI and UV data:

$$M_{p,ls} = \frac{\alpha_{cp}}{(dn/dA)_{cp} * \varepsilon_p} = NM_{monomer} \tag{9}$$

$(dn/dA)_{cp}$ is calculated in a manner analogous to that for the unconjugated protein (Eq. 3). The association state N is determined with Eq. 7. If ε_p is affected by conformational effects resulting from conjugation, this effect should be minor and the association state should be obvious (*i.e.*, round the ratio to the nearest whole number).

Determining the mass of a conjugated protein-polymer

The association state of the conjugated protein was determined above using RI, UV and LS signals. Once the association state is determined, the mass of the conjugated protein, and indirectly the degree of conjugation, will be most conveniently determined using only RI and UV signals. The equation for $(dn/dc)_{cp}$ can be written as:

$$\left(\frac{dn}{dc}\right)_{cp} = \left(\frac{dn}{d\left(\frac{A}{\varepsilon}\right)}\right)_{cp} = \varepsilon_p \frac{M_p}{M_{cp}} \left(\frac{dn}{dA}\right)_{cp} \tag{10}$$

The right-hand side of Eq. 10 is substituted for $(dn/dc)_{cp}$ in Eq. 8 to give:

$$M_{cp} = M_{p,ls} \frac{\varepsilon_p \left(\frac{dn}{dA}\right)_{cp} - \left(\frac{dn}{dc}\right)_p + \left(\frac{dn}{dc}\right)_c}{\left(\frac{dn}{dc}\right)_c} \tag{11}$$

$M_{p,ls}$ is determined by Eq. 9. $(dn/dA)_{cp}$, $(dn/dc)_p$, and $(dn/dc)_c$ are determined by Eq. 3 through experiments on the conjugated protein, the unconjugated protein, and the unconjugated polymer respectively. In cases where similar solvent systems and conjugating polymers have been studied before, literature values of $(dn/dc)_c$ are acceptable.

Case studies

Ribonuclease A (RNase A), protease-free, highly purified, bovine pancreas, was purchased from Calbiochem (La Jolla, CA) and used without further purification. Recombinant *E. coli* derived erythropoietin (EPO) (unglycosylated), chinese hamster ovary cell derived erythropoietin (CHO EPO) (glycosylated), brain-derived neurotrophic factor (BDNF) and PEGylated BDNF (20 kDa PEG) were obtained from Amgen Inc., and used without further purification. Details of pegylation and instrument setup are given in *(9)*. CHO EPO, various PEGylated fractions of monomeric RNAse A, and a dimeric PEGylated BDNF protein were selected to illustrate the utility of the methods described above.

The typical approach was to run SEC – LS/RI/UV first on the unconjugated protein to determine $(dn/dc)_p$ using Eq. 3, followed by a run of the conjugated molecule to determine α (Eq. 8) and $(dn/dA)_{cp}$ (Eq. 3). $(dn/dc)_c$ was determined by flow injection analysis from stock solutions of known concentrations (determined by measured weight in volume) of the conjugating polymer. The molecular mass of the protein, $M_{p,ls}$ is then determined using Eq. 9 and compared to the sequence molecular mass to determine association state. Finally Eq. 11 is used to determine the overall molecular mass of the conjugated protein, which combined with the polypeptide mass and association state will yield the degree of conjugation.

CHO EPO, PEG RNase A, and PEG-BDNF were chromatographed as described above (Figure 1).

$(dn/dc)_p$ of the CHO EPO protein was made for the non-glycosylated form of EPO expressed in *E. coli*. A single injection of 0.25 mg of *E. coli* EPO gave a pure baseline-resolved main peak (data not shown). A $(dn/dc)_p$ value of 0.191 ml/g was determined via Eq. 3. $(dn/dc)_c$ of the carbohydrate in CHO EPO was estimated by flow injection analysis of stock solutions of the carbohydrate β-cyclodextrin (M = 1100 Da) at known concentrations (by weight) in PBS, resulting in $(dn/dc)_c$ of 0.145 ml/g. A single injection of 0.5 mg of CHO EPO was analyzed with LS/UV/RI detection (Fig 1A). Applying Eqs. 8 and 3 resulted in α = 5440 and $(dn/dA)_{cp}$ = 0.230 respectively. The molecular mass of the protein portion was determined with Eq. 9, $M_{p,ls}$ = 18200 Da, which agrees well with the sequence molecular mass of 18,236 Da and confirms that CHO EPO is monomeric. A CHO EPO peptide extinction coefficient ε_p of 1.24 ml/mg/cm was based on dry weight analysis previously published *(11)*. Eq. 11 gives a total molecular mass of 31800 Da for the glycosylated protein, a 42.8% degree of glycosylation (Table I).

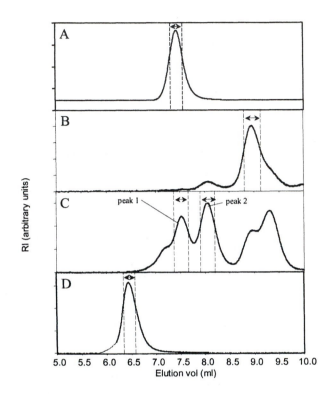

Figure 1. SEC chromatograms of glycosylated and PEGylated proteins. A: CHO EPO (monomer, molecular mass 31,800 Da., 42.8% glycosylated). B: 0.5:1 PEG:RNase A reaction mixture (monomer, molecular mass 18,300, mono-PEGylated with 5000 Da. PEG. C: 5:1 PEG:RNase A reaction mixture (peak 1: monomer, molecular mass 28,600, tri-PEGylated with 5000 Da. PEGs, peak 2: monomer, molecular mass 23,900, di-PEGylated with 5000 Da. PEGs). D: Purified PEG:BDNF (dimer, molecular mass 62,100, each subunit mono-PEGylated with 20,000 Da. PEG). Dashed vertical lines represent regions selected for analysis. (Reproduced from (9) with permission from Elsevier Press.)

Table I Molecular mass, association state, and degree of conjugation of CHO EPO, PEG-RNase A, and PEG-BDN

Protein	$M_{p, ls}$ Eq. 6	$M_{monomer}$ based on sequence	Associatio n state	M_{cp} Eq. 8 ($M_c = M_{cp}$-NM_p)	Degree of glycosylation / PEGylation
CHO EPO Fig. 1A	18200	18236	monomer	31800 (13600)	42.8%
5 kDa. PEG-RNase A Fig. 1B	12800	13682	monomer	18300 (4620)	Mono-PEG
5 kDa. PEG-RNase A Fig. 1C Peak 1	12900	13682	monomer	28600 (14900)	Tri-PEG
5 kDa. PEG-RNase A Fig. 1C Peak 2	12400	13682	monomer	23900 (10200)	Di-PEG
20 kDa. PEG-BDNF Fig. 1D	26000	13513	dimer	62100 (35100)	Mono-PEG (on each monomer)

Data reproduced from *(9)* with permission from Elsevier Press.

Native RNase A was used to determine $(dn/dc)_p$ in PBS with 10% ethanol (PBS-EtOH). Analysis of a single injection of 0.05 mg gave $(dn/dc)_p = 0.197$ ml/g. The $(dn/dc)_c$ of PEG in PBS-EtOH was determined to be 0.133 ml/g by flow injection analysis.

PEGylated RNase A from a solution reacted with a ratio of 0.5:1 PEG:RNase A elutes as shown in Fig. 1B. The extended trailing edge of the peak was not observed in the UV trace and is presumably unreacted PEG. The protein was found to be monomeric with an $M_{p,ls}$ 12800 Da (the sequence mass is 13,682 Da). Eq. 11 yields a total molecular mass of 18300 Da for the PEGylated RNase A peak. Subtracting this value from M_{true} gives the PEG molecular mass of 4620 Da, consistent with a mono-PEGylated product.

RNase A was also PEGylated using a reaction ratio of 5:1 PEG:RNase A. The resulting SEC chromatogram for this reaction (Fig. 1C) indicates multiple PEGylated species. The large peak eluting at 9.3 ml is unreacted PEG based on an injection of PEG alone (data not shown). The results of the molecular mass analyses are given in Table I, and are consistent with a monomeric species that is tri-PEGylated in peak 1 ($M_{cp} = 28600$ Da), and di-PEGylated in peak 2 ($M_{cp} = 23900$ Da).

Native BDNF before conjugation with 20 kDa PEG gave a $(dn/dc)_p$ value equal to 0.204 ml/g in PBS-EtOH. The sequence molecular mass is 13513 Da, and the extinction coefficient is 1.76 ml/mg/cm (determined by amino acid analysis – data not shown). The results for the 20 kDa PEGylated BDNF are presented in Table I and Fig. 1D. $M_{p,ls} = 26000$, which is consistent with a dimeric species based on sequence. $M_{cp} = 62100$, which is consistent with a dimeric species with 2 PEG molecules attached.

Discussion

The discovery that the product $M \bullet (dn/dc)$ is a constant (α) (dependent on a given protein) allowed a simple algebraic solution of the classical light scattering , ultraviolet, and differential refractometry equations to directly give association state. Once association state is known, degree of conjugation can be readily determined. CHO EPO was chosen as a model protein in this study because it is available in pure form and has been well characterized (12-14), allowing direct comparison of previous results to our method. A molecular mass of 31800 Da (Table I) is in good agreement with a previous report of 30390 Da by sedimentation equilibrium (14). Unpurified reaction mixtures comprised of either 0.5:1 or 5:1 ratios of PEG to RNase A resulted in non-baseline resolved chromatograms (Figs 1 B & C). The analysis of 20 kDa PEG-BDNF indicates a dimeric species (Table I), and the total mass data are consistent with addition of 2 PEG molecules per dimer, as expected. Presumably the PEG molecules are attached one per BDNF monomer, but it should be noted the light scattering analysis could not distinguish that situation from one with both PEG molecules on the same BDNF monomer.

It is convenient to use model carbohydrates such as dextran or β-cyclodextrin *(8)* when determining carbohydrate *dn/dc*, due to the difficulties in purifying the glycation moieties from the glycosylated protein. Errors in $(dn/dc)_c$ have a less dramatic effect on M_{cp} than errors in $(dn/dc)_p$. For example, a 5% error in $(dn/dc)_c$ only results in a 1 to 2% error in M_{cp}. Furthermore, Kunitani et. al. demonstrated the similarity of $(dn/dc)_p$ values for various glycation moieties *(3)*.

Previous techniqes required integration of fairly pure baseline-resolved peaks and would not be able to characterize these species without an additional purification step *(4)*. Similarly, sedimentation equilibrium would require purified fractions in order to determine the degree of PEGylation. In contrast, the methods presented here can determine the degree of PEGylation and the association state directly from a single injection of the reaction mixture (Table I). It should be noted that the presence of shoulders or overlapping peaks will influence the derived M_{cp}. This happened with the unpurified RNase A – 5 kDa PEG reaction mixture (Fig. 1B and C). Selecting a narrow region within each peak for analysis minimized the contribution of the adjacent peaks, allowing a more accurate determination of the degree of PEGylation (Fig. 1 and Table I). The techniques outlined here for unpurified proteins should be sufficient in most cases for understanding degree of conjugation and solution association state of conjugated proteins.

References

(1) Fairman, R.; Fenderson, W.; Hail, M. E.; Wu, Y.; Shaw, S. Y. *Anal. Biochem.* **1999**, *270*, 286-295.

(2) Shire, S. J. In *Modern analytical ultracentrifugation*; Schuster, T. M., Laue, T. M., Eds.; Birkhauser: Boston, 1994, pp 261-297.

(3) Kunitani, M.; Kresin, L. *J. Chromatogr.* **1993**, *632*, 19-28.

(4) Kunitani, M.; Dollinger, G.; Johnson, D.; Kresin, L. *J. Chromatogr.* **1991**, *588*, 125-137.

(5) Dollinger, G.; Cunico, B.; Kunitani, M.; Johnson, D.; Jones, R. *J. Chromatogr.* **1992**, *592*, 215-228.

(6) Takagi, T. *J. Chrom.* **1990**, *506*, 409-416.

(7) Hayashi, Y.; Matsui, H.; Takagi, T. *Methods Enzymol.* **1989**, *172*, 514 528.

(8) Wen, J.; Arakawa, T.; Philo, J. *Anal. Biochem.* **1996**, *240*, 155-166.

(9) Kendrick, B. S.; Kerwin, B. A.; Chang, B. S.; Philo, J. S. *Anal. Biochem.* **2001**, *299*, 136-146.

(10) Wyatt, P. J. *Anal. Chim. Acta.* **1993**, *272*, 1-40.

140

(11) Davis, J. M.; Arakawa, T.; Strickland, T. W.; Yphantis, D. A.
Biochemistry **1987**, *26*, 2633-2638.

(12) Rush, R. S.; Derby, P. L.; Smith, D. M.; Merry, C.; Rogers, G.; Rohde, M.
F.; Katta, V. *Anal. Chem.* **1995**, *67*, 1442-1452.

(13) Hokke, C. H.; Bergwerff, A. A.; Van Dedem, G. W.; Kamerling, J. P.;
Vliegenthart, J. F. *Eur. J. Biochem.* **1995**, *228*, 981-1008.

(14) Philo, J. S.; Aoki, K. H.; Arakawa, T.; Narhi, L. O.; Wen, J. *Biochemistry*
1996, *35*, 1681-1691.

(15) Philo, J. S.; Rosenfeld, R.; Arakawa, T.; Wen, J.; Narhi, L. O.
Biochemistry **1993**, *32*, 10812-10818.

(16) Radziejewski, C.; Robinson, R. C.; DiStefano, P. S.; Taylor, J. W.
Biochemistry **1992**, *31*, 4431-4436.

(17) Eisenberg, H.; Josephs, R.; Reisler, E. *Biopolymers* **1977**, *16*, 2773-2783.

(18) Eisenberg, H. *Biochem. Soc. Symp.* **1992**, *58*, 113-125.

Chapter 8

Multiple Detection (Light Scattering, Fluorescence, Refractive Index, and UV) in Size-Exclusion Chromatography of Soluble Glucan Polymers

Wallace H. Yokoyama and Benny E. Knuckles

Western Regional Research Center, Agricultural Research Service, U.S. Department of Agriculture, Albany, CA 94710

β-Glucan ((1→3),(1→4)-β-D-glucan) and methylcellulose are soluble dietary fibers that have physiological properties beneficial to human health. They are often studied for their nutritional properties but they are rarely characterized. We describe the use of size exclusion chromatography and multiple detectors to characterize these polymers: an 18 angle light scattering detector and refractive index detector to determine molecular weight distributions of the polymers, and a postcolumn reaction to form a complex detectable by a fluorescence detector. The fluorescence from the polymer specific complex enables positive identification of the glucan and/or methylcellulose peak of interest when multiple components elute upon chromatography of biological samples. Some applications of this system to characterize the affects of processing and digestion on β-glucans and their relationship to cholesterol lowering; and the effects of colonic fermentation on methylcellulose are described.

Soluble plant polymers have been shown to reduce plasma cholesterol, a risk factor for cardiovascular disease, reduce the glycemic response in the presence of carbohydrate foods, and to improve laxation. Soluble plant polymers are also added to foods and many consumer cleaning aids to increase viscosity. In most nutritional and food studies, characterization of the functional quality or physical characteristics is not reported. There is a need to characterize soluble polymers occuring in studies of plant foods, nutritional supplements, and food additives in order to understand and optimize their beneficial properties. Natural and synthetic glucan polymers are derived from plant sources and are usually difficult to characterize because they have a molecular weight distribution rather than a single molecular weight. Another difficulty is that they are often contained in biological systems such as plant extracts, foods, intestinal lumen, and feces that contain other polysaccharides, protein or nucleic acid polymers with overlapping polymer distributions. Size exclusion chromatography (SEC) is necessary to separate and characterize the polymers. Relative or absolute molecular weight distributions can be determined by retention time of molecular weight standards or multiple angle laser light scattering (MALS) in conjunction with SEC.

We have used SEC-MALS (1) to characterize plant derived glucans such as (1,3), (1,4)- β-D-Glucans (β-Glucans) from cereals and methylcelluloses (MC). Viscosity is usually the property of soluble linear polymers of most interest. Viscosity is directly related to molecular weight. Since carbohydrate polymers do not have good chromophores either refractive index, light scattering or a chemical or enzymatic reaction must be used for their detection. The glucan polymers are usually separated by SEC from plant extracts, foods or biologically derived samples and contain other polymers that may complicate peak identification or overlap with the glucan peak. Glucan polymers containing (1,4) linkages form fluorescent complexes with calcofluor (Fig. 1). The eluting glucan peak can be positively identified by integrating a postcolumn reaction with calcofluor and fluorescent detection. In our research we have used and describe here the use of an 18 angle light scattering detector (photodiode #1 is not available) and a refractive index (RI) detector in order to determine molecular weight distribution and the fluorescence detector to determine the elution of the glucan/MC calcofluor complex. Light scattering instruments with two or more photodiodes can also be used if the Zimm plot is linear. Glucan polymers also form complexes with stronger absorbance at characteristic wavelengths with dyes such as congo red and tinopal CBS-X (2). A diode array detector (DAD) is also often used to detect and characterize proteins or other polymers containing UV absorbing chromophores and could be used to detect glucan-dye complexes.

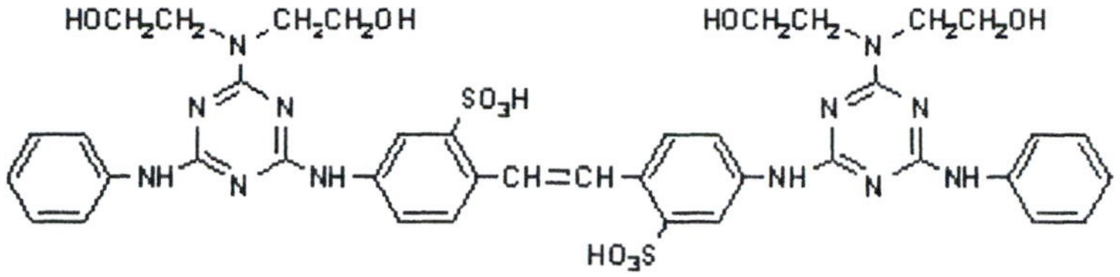

Figure 1. Calcofluor, molecular structure.

Using this multiple detection system we have been able to routinely characterize β-Glucans in complex matrices such as extracts from oat and barley kernels or foods derived from oat and barley, β-Glucans from the intestinal lumen of test animals, and methylcellulose in the intestinal lumen and feces of test animals.

Materials and Methods

Generally solutions for size-exclusion chromatography were prepared by extracting ground oat or barley kernels, freeze-dried foods or biological materials with 0.02% sodium azide (4.5 mL) and filtering (0.45 μm). The polymer components in these solutions (100 μL) were separated by HPSEC. The system included control software, two pumps (both Model 1100, mobile phase and calcofluor reagent), refractive index detector (Model 1100), and fluorescence detector (Model 1046) from Agilent (Palo Alto, CA), Aquagel® columns (3 linear OH + OH-60, Polymer Laboratories, Amherst, MA), and a multiple angle laser light scattering detector (MALLS, Dawn DSP-F, Wyatt Technologies, Santa Barbara, CA). Calcofluor White (CAS 4404-43-7) was purchased from Sigma, St. Louis, MO. The polymer components were separated by elution with 0.02% sodium azide (0.6 mL/min). Positive identification of the MC or β-glucan as the calcofluor complex was achieved by simultaneous fluorometric detection (Ex=415 nm, Em=445 nm) upon introduction of calcofluor reagent (0.1 mg/L in 0.1 N NaOH, 0.7 mL/min) into the eluant stream by a mixing-T following the RI detector. Polymer mass calculated from RI and light scattering data was fitted (first degree, Zimm equation ($K*c/R(q)$)) to a line whose intercept is the molecular weight (g/mol) of the polymer.

Results and Discussion

Instrument configuration

The flow path of the eluting solvent through the columns and four detectors and the mixing with calcofluor reagent is shown in Fig. 2. Signals from the 18 light scattering detectors of the MALS and RI detectors are utilized by the manufacturer's software (ASTRA) to calculate the molar mass of the eluting polymer fraction. A requirement of the calculation is the assumption that the eluting solution is homogeneous. Since the volume of the DAWN DSP detector is extremely small, 37 µl, the eluting fraction contained in that sample volume is considered to be homogeneous. A typical chromatogram of the polymer profile of β-glucan is shown in Fig. 3. The three traces include one angle from the 18 angle light scattering detector (usually the 90°), the RI signal and the fluorescence signal after mixing with the calcofluor reagent. If there are no overlapping signals the RI signal can be used to estimate the amount of eluting polymer since RI is proportional to concentration. The negative peak in the RI signal, due to the difference in RI between the eluting solvent and the extraction solvent (water), is not present if the eluting solvent is used as the extraction solvent.

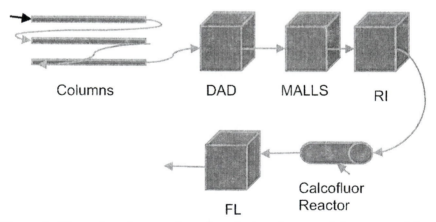

Figure 2. Eluant flow diagram through multiple columns and detector modules.

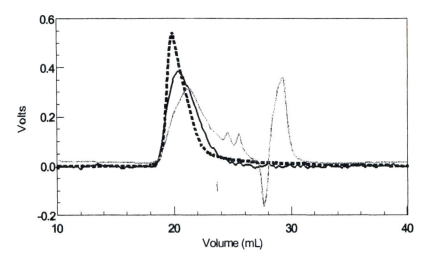

Figure 3. Barley extract. Light scattering, 90° = dashed line, RI = light gray line, Fluorescence (glucan/calcofluor complex) = dark gray. In the RI, the two peaks at 24-26 ml are unknown contaminants and the negative peak is due to the lower RI of water the extraction solvent.

Applications

The use of multiple detectors and the formation of a fluorescent glucan complex can also be used to determine purity of glucan polymers. Coeluting polymer contaminants are an especially difficult problem for carbohydrate polymers since they have no specific wavelength absorbing chromophores, distributions are often unsymmetrical, and elution can be spread out over ten minutes. As an example, the chromatograms of methylcellulose and a commercial β-glucan standard are shown in Figs. 4A,B, respectively. The traces are the RI signal and the fluorescence signal after calcofluor complex formation. Methylcellulose is synthesized from crystalline cellulose and contains only cellulose polymers. The β-glucan standard is extracted from a plant source and may contain other polymers. There is good overlap between the fluorescence and RI signals in the methylcellulose chromatogram, Fig. 4(Upper). The chromatogram of the β-glucan, Fig. 4(Lower), on the other hand, has a stronger RI signal than the fluorescence signal after the peak, suggesting a coeluting polymer impurity.

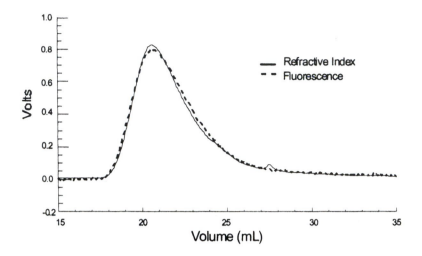

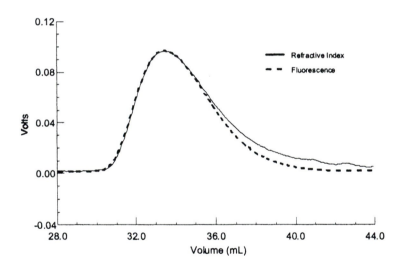

Figure 4. Methylcellulose. Refractive Index and Fluorescence from Methylcellulose/Calcofluor complex (Upper). The peak shape (slopes of up and down for both RI and FL are similar-methylcellulose appears to be pure. A commercial β-glucan standard (Lower). The downward slope suggests that another component may be present in this preparation of glucan.

Oatrim is a commercial preparation of soluble β-glucans from oats containing about 10-11% β-glucan, 11% protein, 5% ash, 2% fat and hydrolyzed starch making up the difference. Oatrim reduces plasma cholesterol from 413±94 mg/dL of the control group fed cellulose fiber to 266±25 mg/dL (3). The oat flour diet also reduced plasma cholesterol to 326±33 mg/dL. Utilizing the SEC and three detector configuration described in Methods, the Mw distribution of β-glucans in Oatrim was compared to oat flour (Table 1). The average Mw of β-glucans in the Oatrim diet was reduced from 1.14×10^6 g/mole to 0.39×10^6 g/mole. In order to determine if this had occurred as a result of mechanical shear of pumping or mixing or enzymatically during the hydrolysis of the starch, oat flour was boiled, stirred and treated with a heat stable alpha amylase. The weight average molecular weight, M_w, of the β-glucan from this oat sample was 1.65×10^6 g/mole, boiling lowered M_w to 1.42×10^6 g/mole, and enzymatic treatment of the boiled oat flour reduced M_w to 0.91×10^6 g/mole. The latter was comparable to the M_w of Oatrim measured at the same time, 0.90×10^6 g/mole. Prior to these studies β-glucan and other natural soluble fibers were often fed to human subjects or test animals without characterization. β-Glucan and other natural polymers are targets of endogenous enzymes that often rapidly reduce the polymer size even in a relative dry form such as flour. Shear during mechanical processes can also reduce polymer size. We have shown in these studies that Mw reduction occurs but did not affect the cholesterol lowering properties of ß-glucans. When β-glucan is isolated from the intestines of hamsters fed oats the molecular weight was found to be approximately 1×10^5 g/mole and much lower than the fed ingredient (unpublished results). This latter observation may explain the lack of effect of the lower average Mw of Oatrim. The average Mw of Oatrim is still ten times higher than the β-glucan products isolated from the stomach and intestines of hamsters and characterized by the triple detector system.

β-Glucanases are not part of the mammalian enzymatic digestion process, however, β-glucan polymers were found to be hydrolyzed in the small intestine (3,4,5) and/or fermented in the large intestine by bacterial enzymes. The physiologically beneficial properties of β-glucans such as cholesterol lowering (6) and glycemic properties (7) are related to polymer length, i.e. viscosity, so that knowledge of polymer stability is important. Methylcellulose polymers are used as a digestive aid, to improve laxation and stool softness. The water-binding and viscous properties of methylcellulose are also related to polymer length. Therefore, the determination of stability to digestive and fermentative hydrolysis is important to understanding how soluble dietary fibers function and to optimize these functions.

Table I. β-glucan Molecular Properties

Component	Enriched Oat Flour	Oatrim-10	Oatrim-5
Molecular Wt, M_w, g/mol	1.136×10^6	0.39×10^6	0.477×10^6 (major) 1.84×10^6 (minor)
Radius, nm	58.9 ± 1.2	17.2 ± 2.0	17.0 ± 1.5 --
Polydispersity, M_w/M_n	1.346 ± 0.084	2.259 ± 0.044	1.020 ± 0.032 (major) 1.545 ± 0.023 (minor)

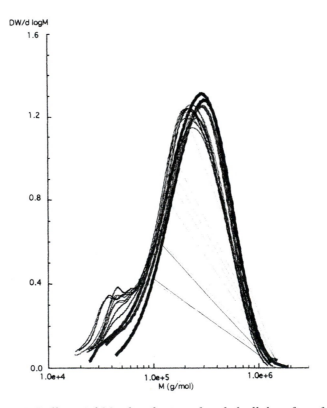

Figure 5. Differential Mw distribution of methylcellulose from feces of ten rats (thin lines) compared to methylcellulose (triplicate, heavy lines) ingredient in feed.

Methylcellulose can be safely consumed since it is not absorbed by the body. However, it can be hydrolyzed in the colon and by fecal bacteria (references cited in (8)) and the physiological activity of the hydrolysates may be compromised. The differential Mw distribution from the extract from the feces of ten hamsters and three methylcellulose ingredient standards are shown in Fig. 5. The degree of hydrolysis was shown to be negligible by comparing the Mw of the starting polymer to the Mw of the polymer in the feces, 3.15×10^5 and 2.82×10^5 g/mol, respectively, and using the equation: $S = M(0)/M(t)-1$. The scission number, S, is calculated to be about 0.1. In other words about one glycosidic bond is broken in every ten methylcellulose polymer molecules. Since each methylcellulose polymer molecule contains approximately 1700 glycosidic bonds, only one bond in about 17,000 bonds are affected. These results indicate that methylcellulose is not significantly hydrolyzed in the colon and retains its water-holding and other beneficial properties. This analysis required the use of the MALS, RI and FL detectors to determine the molecular weight distribution and positively identify the peak of interest.

References

1. Knuckles, B. E.; Yokoyama, W. H.; Chiu, M. M. *Cereal Chem.* **1997,** *74,* 599
2. Wood, P. J.; Fulcher, R. G. *Cereal Chem.* **1978,** *55,* 952.
3. Yokoyama, W. H.; Knuckles, B. E.; Stafford, A. E.; Inglett G. E. *J. Food Sci.* **1998,** *63,* 713.
4. Wood, P. J.; Weisz, J.; Mahn, W. *Cereal Chem.* **1991,** *68,* 530.
5. Sundberg, B.; Wood, P.; Lia, A.; Andersson, H.; Sandberg, A.S.; Hallmans, G.; Aman P. *Am. J. Clin. Nutr.* **1996,** *64,* 878.
6. U.S. Food and Drug Administration. Code of Federal Regulations. Sec. 101.81 Health claims: Soluble fiber from certain foods and risk of coronary heart disease http://vm.cfsan.fda.gov/~lrd/CF101-81.HTML.
7. Bourdon, I.; Yokoyama, W.; Davis, P.; Hudson, C.; Backus, R.; Richter, D.; Knuckles, B; Schneeman B. *Am. J. Clin. Nutr.* **1999,** *69,* 55.
8. Yokoyama, W. H.; Knuckles, B. E.; Davis, P. A.; Daggy, B.P. *J. Agric. Food Chem.* **2002,** *50,* 7726.

Mass Spectrometry
and NMR Spectroscopy

Chapter 9

Size-Exclusion Chromatography/Matrix-Assisted Laser Desorption Ionization and SEC/NMR Techniques for Polymer Characterization

Maurizio S. Montaudo

Istituto per la Chimica e la Tecnologia del Materiali Polimerici, Consiglio Nazionale delle Ricerche, Viale A. Doria, 6–95125 Catania, Italy

Measurements on a series of random copolymers with units of methyl methacrylate (MMA), butyl acrylate (BA), styrene (St), and maleic anhydride (MAH) are performed. A characterization method is used which consists in fractionating the copolymer by size-exclusion chromatography (SEC), collecting 30-40 fractions and then recording both the nuclear magnetic resonance (NMR) spectroscopy and matrix-assisted laser desorption ionization (MALDI) spectra of the fractions. In a successive step, Bivariate Distribution of chain sizes and composition maps are derived from knowledge of the molar mass, weight, and composition of the copolymer fractions.

INTRODUCTION

Size-exclusion chromatography (SEC) is a widely used technique in polymer and copolymer characterization (*1-5*). NMR can be used as a detector for SEC of polymers. SEC and NMR can be connected on-line, using a commercial probe. Work in the field of on-line coupling between liquid chromatography and NMR has been reviewed (*6,7*) and this technique has been applied to homopolymers (*8-10*) and copolymers (*11-14*).

The on-line SEC-NMR technique faces some difficulties. First of all, the signal-to-noise ratio (S/N) of the NMR spectrum is related to the strength of the magnetic field (B) by a nonlinear relationship (S/N scales as B raised to 1.5). Thus, when an NMR spectrometer with a low-field magnet is used, the signal-to-noise ratio of the NMR spectrum is poor. In practice, copolymer analysis requires high-field magnets (600 MHz or higher) and the cost of the SEC apparatus experiences a large increase (by at least one order of magnitude) with respect other SEC assemblies. The NMR probe for on-line coupling is itself a quite sophisticated piece of equipment. As a consequence, double-detector SEC assemblies are more affordable and their use is more widespread than on-line SEC-NMR. Furthermore, although NMR is always able to determine the composition of a copolymer fraction (independent from its molar mass value), it provides reliable molar mass (MM) estimates only up to 10,000-20,000 Daltons. One may try to estimate the molar mass from SEC retention times. Specifically, a mixture of 5-6 or more polymer samples with the same repeats unit, possessing a narrow MM distribution and known mass (the so-called SEC primary standards), is prepared. The mixture is injected into the SEC apparatus and the resulting chromatogram is recorded. Measuring the elution volumes and plotting them against the logarithm of the molar mass, the calibration curve is obtained. However, calibration standards with narrow distribution, known composition, and known molar mass are often not available. For this reason approximate calibration curves are often used. The latter can produce unexpected effects (especially in the case of high conversion samples, which are complex mixtures), as the calibration is logarithmic (i.e. log(M) is used) and the law of propagation of errors in indirect measurements predicts that the $\overline{M}_n$ and $\overline{M}_w$ estimates performed with the use of an inaccurate calibration curve become useless and misleading.

Off-line SEC-NMR does not suffer from the above drawbacks. The signal-to-noise ratio is good when medium-field magnets (200 MHz up to 500 MHz) are employed and thus cost of this SEC apparatus is acceptable, i.e. of the same order of magnitude as double-detector SEC assemblies. Off-line SEC-NMR differs from on-line SEC-NMR, with the former certainly being more time-consuming than the latter. In the off-line experiment SEC fractions are collected, the solvent is evaporated, deuterated solvent is added, and NMR tubes are filled.

This difference between on- and off-line modes can be minimized by reducing the number of fractions. Clearly, the reduction cannot go beyond a certain limit, otherwise it will cause a loss of accuracy in the measurement of copolymer properties.

Mass spectrometry is an emerging technique in polymer characterization *(14)*. It has been demonstrated that one can use a MALDI mass spectrometer and record the MALDI-TOF (time-of-flight) mass spectra of the SEC fractions of homopolymers and copolymers*(15-24)*, since the MALDI-TOF technique possesses extraordinary sensitivity and is able to measure molar masses up to very high values (10^6 Daltons). Furthermore, in order to have a more complete picture of the polymer, one can record both the NMR and the MALDI-TOF spectra of each SEC fraction *(23-24)*.

In this paper, off-line SEC-NMR measurements are performed, along with SEC-MALDI measurements, on random copolymers with units of methyl methacrylate, butyl acrylate, styrene, and maleic anhydride reacted at high conversion. The copolymer is fractionated by SEC, and then fractions are collected. The solvent is evaporated, deuterated solvent is added, and the NMR spectrum is recorded. The polymeric solution is subsequently mixed with the MALDI matrix, the mixture is slowly evaporated, and the MALDI spectrum of the SEC fraction is recorded. The results are employed to derive various copolymer properties such as the composition distribution histogram, which reports the weight fraction of chains with a given composition. The Bivariate Distribution of chain lengths and compositions *(25-27)* is also derived.

CALCULATIONS

In this study copolymers with different structures are analyzed. The equations used to extract sequence distribution and composition data from the copolymer's NMR spectrum change as the structure of the two repeat units in the copolymer changes. The ^{1}H-NMR spectra for random MMA-BA copolymer MB41 yielded an estimate of the copolymer composition *(28)*. The molar fraction of MMA in the copolymer, F_{MMA} is given by:

$$F_{MMA} (x) = 1/[1 + 1.5\ I_{MMA} / I_{BuA}] \tag{1}$$

Where I_{MMA} is the area of the region 4.16-3.92 ppm, corresponding to MMA units, and I_{BuA} the area in the region 3.66-3.51 ppm, corresponding to butyl acrylate units. The variance of Compositional Distribution for MMA, σ^2, was computed from the abundances, I_{MMM}, $I_{MMB+BMM}$, I_{BMB}, of the M-centered triads using the formula:

$$\sigma^2 = \{ Y1^2 + Y2^2 + Y3^2 \}/3 \qquad (2)$$

where $Y1 = (3\ I_{MMM} - k)$, $Y2 = (2\ I_{MMB+BMM} - k)$, $Y3 = (I_{BMB} - k)$, $k = F_{MMA}\ DP_n$, and DP_n is the number-average degree of polymerization. The abundances of the M-centered triads are related to the areas under the MMM, MMB+BMM, and BMB resonances (29) (the co-isotactity factor is 0.40).

The ^{1}H-NMR spectra for random copolymer samples of styrene (St) and maleic anhydride (MAH) labeled as SH91 and SH78 yielded an estimate of the copolymer composition (29) which was determined using the formula:

$$F_{st} = 0.2\ I_{arom}\ (0.5\ I_{alif} - 0.1\ I_{arom})^{-1} \qquad (3)$$

where F_{st} is the molar fraction of styrene in the copolymer, I_{arom} is the area in the region 6-8 ppm corresponding to styrene units, and I_{alif} is the area in the region 1-3 ppm. The overall variance of Compositional Distribution for styrene, σ^2, cannot be easily computed from the areas under the SSS, SSM+MSS, and MSM resonances, I_{SSS}, $I_{SSM+MSS}$, I_{MSM}, since they partially overlap. In order to overcome the cited difficulties distortionless enhancement of polarization transfer (DEPT) NMR subspectra were recorded, a linear combination was taken, and σ^2 was estimated using the formula:

$$\sigma^2 = \{ W1^2 + W2^2 + W3^2 \}/3 \qquad (4)$$

where $W1 = (3\ I_{SSS} - k)$, $W2 = (2I_{SSM+MSS} - k)$, $W3 = (I_{MSM} - k)$, $k = F_{st}\ DP_n$, DP_n is the number-average degree of polymerization, and F_{st} is the molar fraction of styrene in the copolymer.

The assignments for the resonances in the ^{1}H-NMR spectra of random St-MMA copolymers have been reported (30). The molar fraction of MMA in the copolymer, F_{MMA}, and the weight of copolymer in each fraction, w_{copo}, were determined using the formula:

$$F_{MMA}\ (x) = 1/[1 + (3/5)\ I_{arom}\ /\ I_{mo}] \qquad (5)$$

The overall variance of Compositional Distribution for styrene, σ^2, for this copolymer sample was computed from the areas under the styrene-centered triads.

The weight, w_{TMS}, of TMS in the NMR tube is about 0.136 mg and I_{TMS}, the area under the peak at 0 ppm corresponding to TMS, is easily measured. The weight of copolymer in each fraction, w_{copo}, was determined using the formula:

$$w_{copo} = (w_{TMS}\ /\ I_{TMS})\ \tau_1\ I_{copo}\ /\ M_{fra} \qquad (6)$$

where M_{fra} is the number-average molar mass of the fraction, I_{copo} is the sum of the areas of the NMR peaks due to the copolymer, and τ_1 is a dimensionless constant. In the case of St-MAH, $\tau_1 = 1466.2$, which arises from the fact that the St and MAH repeat unit possess 8 and 2 protons, respectively, whereas TMS possesses 12 protons. In the case of St-MMA, $\tau_1 = 916$.

Data can be converted into tabular form where, for each SEC fraction, the first column corresponds to the degree of polymerization (chain size), the second column to the composition (molar fraction of A units), and the third column to the weight in milligrams (or micrograms). Any commercially-available plotting program is able to create a three-dimensional (3D) plot from such a table. These kinds of 3D plots are usually referred to as the Bivariate Distributions. The weight, $W(s, c_A)$, of chains which possess a given size (s) and a given composition (c_A) is related to $I(x)$ (the molar fraction of A_mB_n chains) as follows:

$$W(s, c_A) = \Phi [I(x)] \qquad (7)$$

where Φ takes into account the fact that W is differential in (ds dc_A), whereas I is differential in (dn dm). The compositional distribution histogram reports, instead, the weight $W(c_A)$ of chains which possess given composition (c_A) and is therefore obtained by summation over all chain sizes, namely:

$$W(c_A) = \sum W(s, c_A) \qquad (8)$$

where the summation is over s and it goes from one to infinity.

Equations (1)-(8) were implemented in a computer program called COPOFRAC which is written in Quickbasic and runs on a PC (31). COPOFRAC accepts as input the fractionation conditions, (i.e. the beginning of fraction collection, the calibration curves for the two homopolymers, and the volume of the fraction) and the parameters which describe the heterogeneity of the copolymer sample (i.e. $\overline{M}_n$, $\overline{M}_w$, and the compositional drift) and it gives as output the mass spectrum and the ^{1}H-NMR spectrum of each fraction. COPOFRAC also gives as output the Bivariate Distribution (see equation (7)) and the compositional distribution histogram (see equation (8)).

A special case of SEC of copolymers is two-dimensional chromatography (also referred to as "Chromatographic Cross Fractionation" and "orthogonal chromatography"), where macromolecules having different compositions are first separated in a non-SEC chromatographic column, and a second (SEC) column is used to separate macromolecules having different sizes. Two-dimensional chromatography is the classical method for full copolymer characterization and it allows to generate the Bivariate Distribution of chain sizes and compositions *(32-36)*.

RESULTS AND DISCUSSION

Copolymer sample MB41 is a high (100%) conversion random copolymer of methyl metacrylate (MMA) and butyl acrylate (BA) produced by radical initiation (polymerized in ethylacetate using *tert*-butylperpivalate as initiator). The sample was injected into the SEC apparatus, and about 40 fractions were collected. Several SEC fractions were then subjected to off-line MALDI and NMR analysis, individually. Figure 1 shows the SEC trace along with MALDI-TOF mass spectra for various SEC fractions; it can be seen that the separation process is extremely efficient, as even high molar mass fractions (170000 Daltons) possess narrow distributions.

The mass spectra of these nearly monodisperse samples allowed the computation of reliable values of the molar masses corresponding to the fractions. The log(M) values of the fractions showed a linear correlation with the elution volume of each fraction and allowed the calibration of the SEC trace against MM; the calibrated SEC trace could then be used to compute average molar mass and dispersion of the unfractionated copolymer ($\overline{M}_w = 91000$, $\overline{M}_w = 43000$, $D = \overline{M}_w / \overline{M}_w = 2.1$).

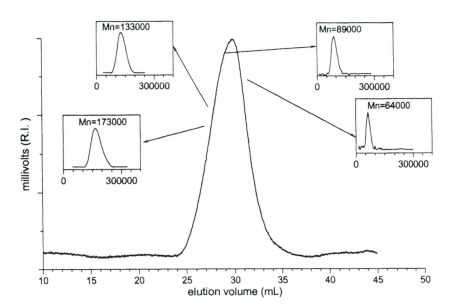

Figure 1. SEC trace for sample MB41. The insets report MALDI –TOF mass spectra of selected fractions

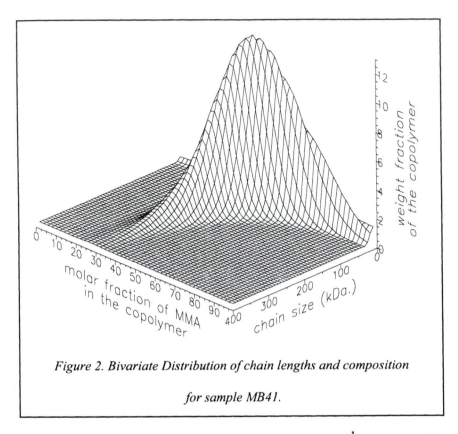

Figure 2. Bivariate Distribution of chain lengths and composition

for sample MB41.

The SEC fractions were also analyzed by 500 MHz ^{1}H-NMR and the signal-to-noise ratio was acceptable for all spectra (it never fell below 13:1). Peaks in the region between 4.2- 3.5 ppm were considered and, more specifically, the region 4.16-3.92 ppm corresponding to MMA units and the region 3.66-3.51 ppm, corresponding to butyl acrylate units. The copolymer composition of each fraction was determined using equation (1). From these data (omitted for brevity) it can be seen that the composition varies and that the fractions taken in the SEC region close to the peak elution volumes possess compositional values close to the average ones (F_{MMA} = 0.41). At higher masses the composition takes values up to 65% BA. At low masses, instead, the macromolecular chains are rich in MMA (about 78%). Peaks belonging to the ^{1}H-NMR spectrum in the 3.66-3.51 ppm region were well resolved. Peaks in this region can be assigned to the M-centered triads and the intensities of the peaks varied when passing from one fraction to another. Inserting peak intensities into equation (2) the variance of the compositional distribution (σ^2) was derived. The variance for fractions 46 and 51 turned out to be slightly lower than the variance for the unfractionated copolymer, but the order of magnitude was the same. This result is not unexpected, as MALDI-TOF data on a series of high-conversion MMA-BA random copolymers indicate that σ^2 for SEC fractions is virtually identical to σ^2 for the unfractionated copolymer *(23)*. The reactivity ratios for the MMA/BA system in ethylacetate are not available. However, it can be assumed that they are identical to the case of toluene *(28)* and thus the predictions of the terminal model can be derived. Monomer MMA is consumed more quickly than BA and thus, at later stages of the reaction (i.e. at high monomer conversion), the chains produced are rich in BA.

The computer program COPOFRAC was used to generate the Bivariate Distribution of chain sizes and compositions for sample MB41 and the result is shown in Figure 2. It is apparent that the sample possesses a high molar mass tail which is made of BA-rich chains and the terminal model allows us to tell that the tail was produced at later stages of the reaction (i.e. at high monomer conversion).

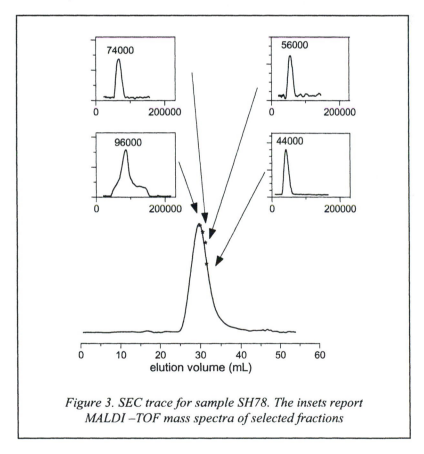

Figure 3. SEC trace for sample SH78. The insets report MALDI –TOF mass spectra of selected fractions

Copolymer samples SH91 and SH78 are random copolymer samples of styrene (St) and maleic anhydride (MAH) obtained at high conversion from solution polymerization using AIBN (*N,N'*-azo-bisisobutyronitrile) as the initiator. The ^{13}C-125 MHz NMR spectrum of samples SH91 and SH78 (not shown) does not yield sequence information because the signals due to methine and methylene carbons partially overlap. However, sequence information can be obtained from DEPT experiments, by recording the spectra at $\pi/4$ and at $3\pi/4$ and then combining them. For sample SH91, the areas under the SSS, SSM+MSS, and MSM resonances are 0.32, 0.50 and 0.18, respectively. In a similar manner, the areas under the SSS, SSM+MSS, and MSM resonances for sample SH91 were 0.39, 0.47 and 0.13, respectively. Inserting the three values for the SSS, SSM+MSS, and MSM resonances in equation (4), the overall variance of Compositional Distribution was found to be $\sigma^2 = 0.0844$ for sample SH78 and $\sigma^2 = 0.0676$ for sample SH91. The theoretical variances for SH78 and SH91 are $\sigma^2 = 0.015$ and $\sigma^2 = 0.0097$, respectively, which are lower than the values calculated using equation (4). Knowledge of the average composition and of the variance about the average may give some hints of the abundance of chains which possess a composition that differs from the average, although this is not sufficient to reconstruct the compositional distribution. For this reason it is necessary to adopt a different approach, based on fractionation. Copolymers SH91 and SH78 were fractionated by SEC and 40 fractions were collected for each copolymer.

Figure 3 reports the SEC trace for SH78, along with MALDI-TOF spectra of selected fractions. It can be seen that the fractions yield excellent MALDI-TOF spectra with narrow distributions (the polydispersity index, D, is often smaller than 1.1) up to high molar masses (up to 96 KDa). The copolymer fractions were also analyzed by ^{1}H-NMR. The copolymer composition of each fraction was determined by measuring the area of the regions 6-8 ppm and 1-3 ppm and by combining them as described in equation (3). The composition values obtained by this procedure imply that the molar fraction of styrene in both copolymer samples varies as the mass of the chain grows. In the case of sample SH91 (data omitted for brevity) the average molar fraction of styrene is in the range 0.58-0.99. Sample SH78 was found to possess a more limited compositional heterogeneity, with the average molar fraction of styrene is in the range 0.53-0.89. The SEC data indicate that the copolymer fractions have a composition close to alternating (50%) for a molar mass of about 17-19 KDa. This molar mass corresponds to the kinetic chain length of the alternating copolymer and indicates that lower and higher molar masses are subjected to a higher percent of styrene inclusion.

The ^{1}H-NMR data was then used to measure the amount of copolymer contained in each fraction, by measuring I_{TMS} (the area under the peak at 0 ppm, corresponding to tetramethylsilane) and combining this value with the quantities I_{arom} and I_{alif} as defined in equation (6) (data omitted for brevity). To measure the amount of copolymer one may rely on a "traditional" SEC detector such as a differential refractometer and assume that the detector's response reflects the amount of copolymer at each elution slice. There are some discrepancies between the cited refractive index (R.I.) values and the values obtained by ^{1}H-NMR. The discrepancies between the two measurements are often small; however, in some cases they become large and cannot be neglected. For instance, the two values for the amount of copolymer in fraction 36 are 0.017 (R.I.) and 0.026 (^{1}H-NMR), which implies a difference larger than 30%. This difference is due to the fact that we are dealing with compositionally heterogeneous copolymers; the refractometer's response to styrene units and to maleic anhydride units may be different and thus unreliable. A UV detector is commonly added to the apparatus and used to correct for the different values of the specific refractive index increment (dn/dc) for the two monomers. However, the time-lag estimation between the two detectors is cumbersome *(1-5)*.

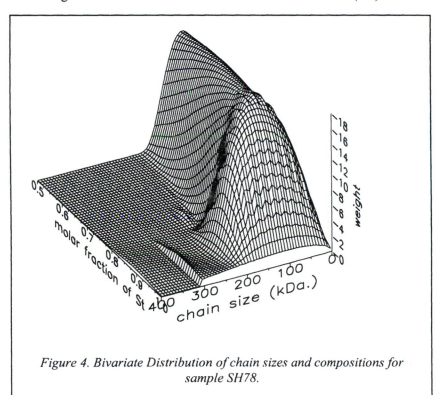

Figure 4. Bivariate Distribution of chain sizes and compositions for sample SH78.

The computer program COPOFRAC was used to generate the Bivariate Distribution of chain sizes and compositions for sample SH78 and the result is shown in Figure 4. The surface displays a single maximum, which is located at F_{St}= 0.85 and molar mass 150 KDa. These results can be understood as due to the MAH monomer being consumed in the first part of the copolymer forming reaction.

The average molar mass and dispersion of the unfractionated copolymers were then computed. The software for such calculations needs the calibration curves for the sample and the abundances. The calibration data were taken from the MALDI-TOF analysis of each fraction. The result was $\overline{M}n$ = 151000, $\overline{M}w$ = 345000, D=2.28. Copolymer sample M30 is a random copolymer with units of styrene (St) and methyl metacrylate (MMA), obtained at high conversion. The copolymer was injected in the SEC apparatus, the SEC fractions were collected and analyzed by ^{1}H- NMR and MALDI. The signal-to-noise ratio in the 200 MHz ^{1}H-NMR spectrum of fraction 34 (not shown) is high, demonstrating that off-line SEC-NMR can be performed using medium-low field magnets (200 MHz), a very attractive feature indeed!

From SEC-NMR data it can be seen that the composition varies dramatically as the molar mass increases. Specifically, the molar fraction of styrene is very high at low masses (14.5 KDa), then falls steadily (in an almost linear manner) until it reaches a value of 0.53 at high mass. SEC fractions of M30 yielded MALDI-TOF mass spectra with a peculiar feature, namely strong signals due to doubly-charged ions. The calibration data were used to compute the molar mass averages, which were $\overline{M}n$ = 99000, $\overline{M}w$ = 240000, D=2.4.

The amount of copolymer in each fraction was computed using equation (6). Figure 6 shows the SEC traces of the M30 sample: before correction (full line) and after correction (dotted line). From the inspection of the Figure, it can be seen that the two SEC chromatograms are quite different and that the full line takes its highest value about 1 mL earlier than the dotted line. In order to show the effects of this difference, the SEC chromatogram recorded using the RI detector was used to compute $\overline{M}n$ and $\overline{M}w$. The result was biased (both averages were underestimated, as expected), corresponding to a 33% error in the determination of $\overline{M}n$ and $\overline{M}w$.

In the case of sample M30 four different fractionation experiments were performed and 50 fractions of 0.2 mL, 25 fractions of 0.4 mL, 15 fractions of 0.8 mL, 15 fractions of 1 mL were collected. The goal of these experiments is to decrease the time for analysis by reducing the number of fractions. In other words to find the optimal conditions, namely, the largest volume of the fraction where the aforementioned loss of accuracy is small, and to collect experimental data which can be used to test the model for copolymers obtained by SEC fractionation described in the theoretical section.

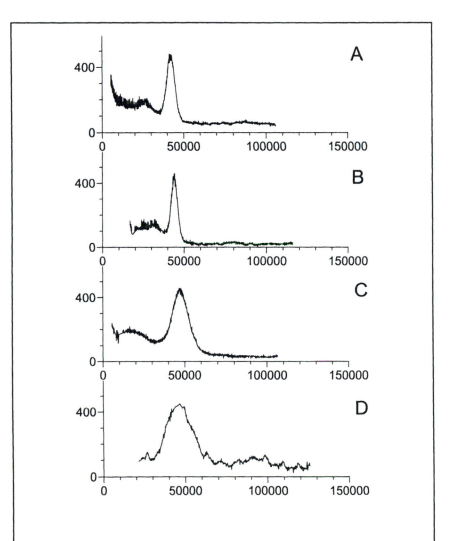

Figure 5. MALDI-TOF mass spectra of SEC fractions of sample M30 collected in four different experiments. The volume of the fraction is 0.2 ml (A) , 0.4 ml (B) 0.8 ml (C), 1.0 ml (D).

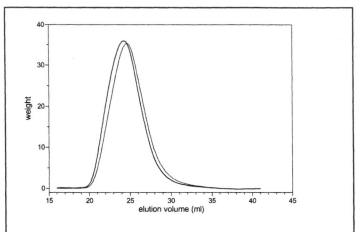

Figure 6. SEC traces of the M30 sample: before correction (full line) after correction (dotted line).

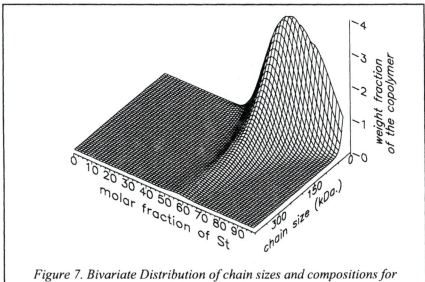

Figure 7. Bivariate Distribution of chain sizes and compositions for sample M30.

Figures 5A-D show the MALDI-TOF mass spectra of the fractions collected around 29 mL. The four spectra are bell-shaped, the tallest molar mass is about 46000 Da, and it is quite apparent that the spectrum becomes broader with increasing fraction volume. The MALDI-TOF mass spectra of the fractions collected around 30 mL (not shown) are bell-shaped also and the highest molar mass is about 26000 Da. It is apparent that when V_1 (the volume of the fraction) is larger than 0.6 mL, D becomes unacceptably large. On the other hand, when V_1 is smaller than 0.6 mL, D is acceptable and thus these represent the optimal conditions.

The program COPOFRAC was used to generate the Bivariate Distribution of chain sizes and compositions for sample M30 and the result is shown in Figure 7. The surface displays a single maximum, located at $F_{St}= 0.75$ and molar mass 100 KDa. The surface is not perfectly symmetrical since, at low masses, the styrene-rich wing is taller than the MMA-rich wing. The absence of symmetry denotes high conversion, since samples reacted at low monomer conversion possess a symmetrical surface *(25-27)*.

CONCLUSIONS

A new method for full copolymer characterization is presented which employs off-line SEC-NMR and SEC-MALDI. A series of examples of application to copolymers reacted at high conversion were discussed. Data were shown demonstrating that off-line SEC-NMR and SEC-MALDI can be performed using medium-low field magnets (200 MHz), a very attractive feature as it allows the use of affordable equipment. The new method represents a valid, less time-consuming alternative to cross-fractionation. While it may appear necessary to collect a large number of fractions for good results, it is shown here that this is not the case and that a relatively small number of fractions is enough to characterize a typical industrial copolymer sample.

REFERENCES

1   Dawkins JV in Allen G, Bevington J, (eds) " Comprehensive Polymer Science ", Pergamon, Oxford, 1989, Vol 1, chapter 12
2   Gores P , Kilz P in Provder T (ed) "Chromatography of polymers", ACS Symp Ser 521, ACS Publ, Washington, 1993, chapter 10
3   Protschka M, Dubin PL, (eds), "Strategies in Size Exclusion Chromatography", ACS Symp Ser 635, ACS Publ, Washington, 1993, chapter 15
4   Cooper AR in Bark LS, Allen NS, (eds), "Analysis of Polymers Systems", Applied Science, London 1982, chapter 8

166

5   Trathnigg B, *J. Chrom.,* 507, 552, 1991, see also Trathnigg B, Yan X, *Chromatographia,* 467, 33, 1992

6   Albert K, *J. Chromatogr.* **1995,** A703, 123,

7   Korhammer S, Bernreuther A, *Fresenius J. Anal. Chem.,* **1996** 354, 131,

8   Ute K, Niimi R, Hongo H, Hatada K,*Polymer J.,* **1998,** 30, 439,

9   Hatada K, Ute K, Okamoto Y, Imanari M, Fujii N, *Polymer Bull.,* **1988,** 20, 317,

10 Hatada K, Ute K, Kitayama T, Nisimura T, Kashiyama M, *Polymer Bull.,* **1990,** 22, 549,

11 Ute K, Niimi R, Hatada K, Kolbert AC, *Int. J. Pol. Anal. Char.,* **1998,**30, 888,

12 Hatada K, Ute K, Kitayama T, Yamamoto M, Nisimura T, Kashiyama M, *Polymer Bull,* **1989,**21, 489,

13 Kramer I, Pasch H, Handel H, Albert K, *Macromol. Chem. Phys.,* **1999,**200, 1734,

14 Montaudo G, Lattimer R.(eds) " Mass spectrometry of polymers " CRC Press, Boca Raton (2002)

15 Montaudo G, Garozzo D, Montaudo M.S, Puglisi C, Samperi F, *Macromolecules,* **1995,**28, 7983,

16 Montaudo G, Montaudo M.S, Puglisi C, Samperi F, *Rapid Commun. Mass Spectrom.,* **1995,**9, 1158,

17 Montaudo G, Montaudo M.S, Puglisi C, Samperi F, *Int. J. Polym. Anal. & Charact.,* **1997,** 3, 177,

18 Montaudo M.S, Puglisi C, Samperi F, Montaudo G, *Macromolecules,* **1998,**31, 3839,

19 Carroccio S, Rizzarelli P, Puglisi C, *Rapid Commun. Mass Spectrom.,* 2000, 14, 1513,

20 Lou X, Van Dongen JLJ, Meijer EW, *J. Chromat.,* **2000,** A896, 19,

21 Esser, E, Keil C, Braun D, Montag P, Pasch H, *Polymer,* **2000,** 41, 4039,

22 Montaudo M.S, Puglisi C, Samperi F, Montaudo G, *Rapid Commun. Mass Spectrom,* **1998,**12, 519,

23 Montaudo MS, Montaudo G, *Macromolecules,* **1999,**32, 7015,

24 Montaudo MS, *Macromolecules,* **2001,** 34, 2792,

25 Murphy RE, Schure MR, Foley JP, *Anal. Chem.,* **1998,** 70,1585,

26 Ogawa T, *J. Appl. Pol. Sci.,* **1979,** 23, 3515,

27 Kuchanov SI, *Adv. Polym. Sci,* **2001,** 52, 157,

28 Tobita H, *Polymer,* 1998, 39, 2367,

29 Aerdts AM, German AL, Van der Velden GPM, *Magnet. Reson. Chem,* 1994, S80, 32,

30 Barron PF, Hill DJH, ODonnell JH, Osullivan PW, *Macromolecules* **1967,** *17,* 1984,

31 Aerdts A, DeHaan JW, German AL, *Macromolecules,* **1993,** 26, 1965,

32 Mori S. *Anal. Chem.*, **1988,** 60, 1125,

33 Glockner G, Koningsveld R, *Makromol. Chem., Rapid Comm.*, **1983,** 4, 529,

34 Ogawa T, Sakai M, *J. Polym. Sci., Polym Phys.*, **1981,**19, 1377,

35 Balke ST, Patel RD, *J. Polym. Sci., Polym. Lett.*, **1980,**18, 453,

36 Schunk T.C. *J. Chromatogr. A*, **1994,** 661, 215

Chapter 10

Size-Exclusion Chromatography–Inductively Coupled Plasma Mass Spectrometry

An Important Analytical Tool for Elemental Speciation in Environmental and Biological Samples

Baki B. M. Sadi[1], Anne P. Vonderheide[2], J. Sabine Becker[2], and Joseph A. Caruso[1,*]

[1]Department of Chemistry, University of Cincinnati, Cincinnati, OH 25221
[2]Central Department of Analytical Chemistry, Research Center Jülich, Jülich, Germany D–52425

The coupling of size-exclusion chromatography (SEC) to inductively coupled plasma mass spectrometric (ICP-MS) detection has proven to be an exceptional analytical strategy. SEC promotes a size-based separation of macromolecules while ICP-MS yields elemental profiles. In this review, particular considerations of the interfacing of these two techniques are detailed. Furthermore, applications of this coupled technique are discussed. Specifically, the use of SEC-ICP-MS in the growing area of elemental speciation studies is of great importance because mobility, bioavailability and toxicity of the different physico-chemical forms of the elements largely depends on their elemental distribution in the different size fractions of the macromolecules. Additionally, SEC-ICP-MS has been found to be a great use in the analysis of biological macromolecules, most often proteins, as research efforts extend to the elucidation of reaction mechanisms and evaluation of binding capacities of different elements with biological entities.

Introduction

Size-exclusion chromatography is a variation of high performance liquid chromatography in which the separation is based on the differential degree of permeation of molecules of different sizes and shapes with respect to the pore size and packing geometry of the stationary phase. Larger molecules take shorter paths through the stationary phase and elute earlier. Because separation is due to the differential permeation of molecules from the moving interstitial liquid (mobile phase) into the stationary liquid inside the pore structure of the packing, interactions between the analyte and the stationary phase are undesirable. Furthermore because retention is based on the physical impedance of the analytes (entropy controlled) rather than chemical interactions (enthalpy controlled), the mobile phase does not play a critical role but should be a strong solvent for the sample. Salts of neutral electrolytes, organic modifiers and buffers are commonly added to the mobile phase to avoid the non ideal behavior (electrostatic and hydrophobic interactions) and to maintain the structural and conformational integrity of the analyte in solution. In general, this technique is used for analytes with molecular weights greater than 2000 such as proteins and polymers as well as other natural and synthetic macromolecules. It is noteworthy that if the retention behavior is correlated to molecular mass, it is possible to approximate the mass of the unknown species. Because the solvent molecules are normally the smallest molecules, the solvent peak is usually the last to elute, predicting the end of the chromatographic run.

The inductively coupled plasma is a high temperature plasma discharge that represents a very efficient way in which the sample can be dissociated into its constituent atoms and subsequently ionized to generate positively charged ions. When coupled to a suitable mass analyzer such as a quadrupole, time of flight or a sector field instrument, it becomes one of the most powerful liquid chromatographic detectors, owing to its unique combination of sensitivity, selectivity, wide linear dynamic range, nearly interference-free operation and multielement capabilities. Detection limits are in the part-per-trillion range for most of the elements of the periodic table. Additionally, the plasma produces mainly singly charged ions that can be identified according to their isotopic abundance. Therefore, in principle, inductively coupled plasma mass spectrometry (ICP-MS) becomes an excellent analytical tool for trace elemental speciation in macromolecules separated by size-exclusion chromatography.

Considerations in the coupling of SEC to ICP-MS

In general, the interfacing of liquid chromatography to ICP-MS is relatively straightforward. The effluent from the column is carried, via

PolyEtherEtherKetone (PEEK) tubing, directly into the nebulizer. A carrier and a make-up gas are brought in at this point to allow for the effluent to be transformed into an aerosol. A spray chamber is used to sort the aerosol droplets according to their size and only the smaller droplets are allowed to pass into the plasma.

Overall, the coupling of SEC to ICP-MS detection is facilitated by the lack of involvement of the mobile phase with regard to the chromatographic separation. However, there are some specific mobile phase considerations with the employment of ICP-MS detection. For example, the use of organic solvents should be kept minimum with ICP-MS because they can adversely affect the stability of the plasma and also deposit carbon on the sampling and skimmer cones.[1] These phenomena can result in increased noise and varying responses; a cooled spray chamber may be employed to minimize the amount of solvent going into the plasma. The ionic strength of the mobile phase buffer must also be considered as it plays an important role in maximizing the molecular sieving mechanism and minimizing the ionic and hydrophobic interactions between the sample and the column packing materials.[2] An increase in buffer salt concentration may help in minimizing the non ideal SEC behavior such as electrostatic[3] and hydrophobic[4] interactions due to charged groups in the stationary phase. However, the high salt concentration may denature the organo-metal complexes.[5] Additionally, the salt content in the mobile phase should be kept to a minimum (<1% total dissolved solids) to prevent clogging of the nebulizer and unnecessary wear on the sampler and skimmer cones. Finally, as in all SEC separations, drastic changes in the pH of the mobile phase should be avoided to prevent precipitation or adsorption of macromolecules within the column. For example, in the analysis of metals complexed to metallothioneins, control of the pH of the mobile phase is crucial since the complexation equilibria between metallothioneins and metals are strongly pH dependent.[6]

In addition to mobile phase constituents, sample preparation is a second area that must be given special consideration in SEC-ICP-MS. SEC does make possible the adaptation of the pH and ionic strength of the mobile phase to the properties of the sample to preserve the nature of the species.[7] However, the species of interest must first be efficiently extracted from the matrix without changes in natural structure in addition to the fact that the native binding of the elements to the macromolecules must not be altered. To such an end, pH must again be carefully controlled to prevent unwanted precipitation or alteration. Mild procedures are required and, in some cases, concentration techniques may be necessary due to the low levels of some species. Preconcentration techniques must be closely monitored as they may lead to the alteration of species equilibria.[5]

A final issue relates to the lack of commercially-available standard compounds.[8] For those that are available, identification based upon retention times matched to known standards is common protocol. If standards are not available, the only information to be gained is an approximate molecular mass of the analyte species estimated on the basis of the elution volume of the complex. Unidentified peaks may also be elucidated by other means such as mass spectrometry coupled with ionization sources that are gentler than the inductively-coupled plasma.[9,10]

Applications of SEC-ICP-MS in Elemental Speciation in Environmental Samples

When coupled to a very sensitive and element specific detection system, size-exclusion chromatography can be a very useful analytical tool to study elemental speciation in the molecular size/weight domain of natural or synthetic macromolecules in environmental samples. These types of studies are of great importance because mobility, bioavailability and toxicity of the different physico-chemical forms of the elements largely depend on their elemental distribution in the different size fractions of the macromolecules. Some of the interesting recent applications of SEC-ICP-MS in speciation analysis are outlined below.

Natural organic matter is rich in humic substances that play a crucial role in transport, mobility and bioaccessibility of toxic trace metals as well as hydrophobic organic compounds (such as halogenated pesticides and polychlorinated hydrocarbons) to different environmental compartments. Size-exclusion chromatography coupled to ICP-MS proved the viability of such investigations by determining metal concentration associated with different size fractions of natural organic matter of a bog lake water.[11] Furthermore, size-exclusion chromatography inductively coupled plasma mass spectrometry using isotope dilution technique (SEC-ICP-IDMS) has been employed to investigate halogen species bound to the humic substances in seepage water samples from soil. This study indicated that the speciation status of chlorine, bromine and iodine in humic substances are not identical.[10] The same SEC-ICP-IDMS system has been used to study the heavy metal complexes with humic substances.[10, 11] In seepage water samples from soil, copper, zinc, and molybdenum were found to form complexes with similar size fractions of humic substances. On the contrary, humic complexes of these metal ions show a different distribution pattern in a water sample from a sewage plant. Similar techniques have also been utilized to investigate the kinetic stability of chromium and copper complexes with humic substances.[12]

Haraguchi *et al.*[12] investigated the speciation of yttrium and lanthanides in natural waters by using SEC-ICP-MS. UV absorption detection was also employed and results showed the detection of yttrium and the lanthanide elements La, Ce and Pr corresponded to the position of large organic molecules. Additionally, water samples were taken at various depths and the light lanthanide elements (La, Ce, Pr) in the dissolved form provided lower concentrations in the middle part of the water column examined, whereas Yb and Lu (heavy lanthanide elements) showed almost constant concentrations through the water column.

There has been a growing interest for aluminum speciation in forest soil due to the importance of different aluminum species in forest ecosystems. Hils *et al.* studied the less phytotoxic organic aluminum species in the percolating water of the forest floor by SEC-ICP-MS.[13] These species are important due to the fact that aluminum, by its coordination to phenolic groups of low molecular weight organic components, may hinder their reaction to high molecular humic substances, and the sorption of the organic components to the forest soil (humus disintegration).

Management of solid organic waste generated by anthropogenic and industrial activities has been a matter of recent concern. As an agricultural supplement for boosting the soil nutritional values, composting is found to be a very efficient method to manage these wastes in an economic as well as environmentally-friendly manner. However, the quality of the compost in terms of its nutritional and toxic element content should be determined before application for agricultural purpose. ICP-MS was used as a detection system for a high performance size-exclusion chromatographic separation to study the elemental binding to different molecular weight fractions of humic substances in compost extract obtained from urban solid waste.[14] The elements investigated (Cr, Mn, Co, Ni, Cu, Zn, As, Mo, Cd, Pb, Th, and U) were found to have differential binding preference for different size fractions of the humic substances. The effect of the metal ions on the molecular weight distribution of humic substances was studied in a similar study with SEC-ICP-MS.[15] This study indicated that both bridging between the small molecules and complexation/chelation by individual molecules are involved in metal ion binding to humic substances.

In the area of food analysis, multi-elemental speciation of tea infusion was studied by SEC-ICP-MS by Odegard and Lund.[16] Fourteen elements were monitored, but only Fe, Ni, Cu, Sr, Ba, Pb and Al were found to be associated with organic complexes. Further experiments with cation-exchange chromatography led to the conclusion that the metal-binding organic ligands were large polyphenolic compounds.

Foodstuffs of plant origin contain significant concentrations of polysaccharides of which the potentially negatively charged oxygen constituents can bind metal cations electrostatically or chelate them via polyhydroxy groups. Szpunar *et. al.* studied the speciation analysis of biomolecular complexes of lead in wine by SEC-ICP-MS.[17] Their study showed that the dominant species that accounts for 40-95% of the lead in wine was a complex formed with the dimer of a pectic polysaccharide, rhamnogalacturonan II. Both boron and lead were monitored, as the presence of boron is necessary in order that a metal complex can be formed with RG-II.

Metal-carbohydrate complexes in fruit and vegetable extracts were studied by the same group using SEC-ICP-MS.[18] Analysis of the water-soluble fraction of apple and carrot samples showed the identification of a high molar mass polysaccharide fraction (>50 kDa) containing Pb, Ba, Sr, Ce and B, in contrast to other metals (Zn, Cu and Mg), which eluted as complexes with low molar mass non-carbohydrate compounds. However, the majority of the metal-carbohydrate complexes were located in the solid water-insoluble fraction. An extraction procedure that utilized pectinolytic enzymes was employed to release this species into the aqueous phase and the metal-binding carbohydrate component was subsequently identified as the dimer of rhamnogalacturonan-II, a pectic polysaccharide present in plant cell walls.

Applications of SEC-ICP-MS to Biological Macromolecules

Size-exclusion chromatography is widely used in protein purification, although it has low resolving power compared to other techniques typically employed in the separation of proteins, such as two-dimensional polyacrylamide gel electrophoresis (2D-PAGE).[19] Given the large number of proteins expressed by even a simple living organism, the use of SEC for their separation is not feasible. However, the coupling of an element-specific detector facilitates investigation of particular elements either incorporated into or simply bound to proteins. In this vein, a number of studies have been performed to elucidate reaction mechanisms and to evaluate binding capacities of different elements with biological entities. Furthermore, the high elemental sensitivity of ICP-MS is advantageous in the detection of such proteins expressed at very low levels. Finally, predictable separation mechanisms in conjunction with high loading capacities further serve to enhance the attractiveness of the coupling of SEC to ICP-MS. Many applications of this universal separation method coupled to ICP-MS detection to both proteins and other biological macromolecules have been recently documented and are discussed in detail below.

Plants

Research is increasing in the ecotoxicity of heavy metals, their pathways in the ecosystem and their metabolism by living organisms. In this pursuit, Vacchina and coworkers[20] utilized SEC for the separation of cadmium phytochelatins (PCs). Phytochelatins are oligopeptides (up to 11 amino acids) that are known to be biosynthesized by plants exposed to metal stress and this method was applied to the investigation of the speciation of Cd in cytosols of plant tissues and plant cell cultures. Because of the high cysteine content, phytochelatins are able to chelate heavy metals, thereby reducing the concentration of free metal ions in the cytosol. Standards of the PCs were synthesized by polymerization of glutathione in the presence of phytochelatin synthase; the resulting mixture of oligopeptides was then isolated by precipitation with cadmium. Due to the size of the PC peptides, a SEC column (Superdex Peptide HR) designed for the separation of amino acids and oligopeptides was investigated.

Additional metals and their binding properties with respect to phytochelatins were investigated by Leopold *et al.* by SEC-ICP-MS.[21] The induction of phytochelatins and the binding of heavy metals to the complexes were investigated by exposure of cell cultures to different concentrations of Cd, Cu, Pb and Zn. An *in vitro* heavy metal saturation assay and *in vivo* stress experiments were performed in order to characterize the binding affinity and binding stability of these compounds in *Silene vulgaris* cell cultures. Results showed that phytochelatins may be induced by several metals that may not necessarily be incorporated into them. Further, copper was shown to bind most stably to PCs and that a ten-fold excess of Cd is not sufficient to remove it from its binding site.

Klueppel *et al.* studied the metabolization of platinum in cultivated grass samples grown with and without the treatment of a Pt containing solution.[22] Conclusions on the binding partners of Pt were drawn from multi-element determinations of selected elements that co-eluted with platinum during the SEC separation. For example, carbon was used as an indicator element and UV detection was used prior to ICP-MS. This allowed for distinguishing between Pt-binding with carbonates/small organic molecules as opposed to large conjugated organic systems. Sulfur was also used as an indicator element in an effort to determine platinum binding to the sulfur amino acids or peptides used by the plants for metal complexation.

Several edible plants or plant fruits have been studied specifically in an effort to determine protein complexes of minerals and trace elements. Koplik *et al.* used SEC-ICP-MS to study phosphorus and other trace elements in soybean flour and common white bean seeds.[23] Results exhibited the molecular weight distribution of elemental species as well as the relationships among the binding forms of the individual elements. In the area of quantitation, Mestek and coworkers examined the use of both external calibration and isotope dilution of copper and zinc species in legume seed extracts.[24]

Kannamkumarath and coworkers studied selenium bound to the proteins of Brazil nuts by SEC-ICP-MS.[25] The proteins were isolated from the nut samples by dissolution in 0.1 M NaOH and subsequently precipitated with acetone. They were then dissolved in phosphate buffer at pH = 7.5 in preparation for SEC separation. This protocol was followed in an effort to determine the percentages of selenium firmly-bound to proteins (in the form of seleno amino acids incorporated into proteins during their synthesis) as opposed to the weakly-bound fraction (weakly-associated through selenodisulfides or methylselenylsulfides formed during post-translational modifications of the proteins). The weakly-bound protein fraction was determined in a second SEC run by the addition of β-mercaptoethanol to the mobile phase as a reducing agent, as the weakly-bound fraction can be easily reduced to low molecular weight selenium compounds.

Mounicou et al.[26] employed SEC-ICP-MS to investigate to binding of lead and cadmium within the biological matrix of the cocoa bean. Interest in these particular metals arose as a result of public interest with regard to their toxic properties. Several extraction methods were investigated in terms of both extraction efficiency as well as successful transfer of the intact metal species into the aqueous phase. An analytical strategy was applied to effect the separation of four types of biological constituents to which the metals potentially may be associated and these included water-soluble entities, water-insoluble protein complexes, water-insoluble polysaccharide complexes and bioavailable metal complexes. The different extraction approaches coupled with species-selective monitoring of the extracted compounds resulted in a more comprehensive understanding of the bioavailability of lead and cadmium in the cocoa product. Results showed that over 70% of the Pb and Cd is bound in very stable complexes that are resistant to gastrointestinal conditions, such as microfibers of crystalline cellulose. An additional experiment with the proteinaceous fraction showed that the molar excess of the Cd- and Pb-binding ligands is very high with respect to the metal available in the metal extract, indicating the lack of biosynthesis of specific ligands, such as phytochelatins.

Animals

Metallothioneins are specific cysteine-rich proteins in mammals capable of binding high amounts of various metal ions in the protein molecules. These molecules intervene in the metabolism and homoeostatic control of a number of essential elements (Zn, Cu) and are involved in the detoxification of toxic trace elements (Cd, Hg).[27] Metal complexes with bio-induced ligands are distinguished from those that are naturally present in the cytosol by comparing the SEC-ICP-MS chromatograms from a control and from an exposed animal. These proteins, as well as the metallothionein-like proteins (MLPs) of marine invertebrates, can therefore be used as biomarkers of heavy metal contamination. In early work concerning the speciation of metallothioneins, Tessier *et al.*, used SEC-ICP-MS to investigate potential genetic induction of metallothionein biosynthesis of zebra mussels (*Dreissena polymorpha)* upon exposure to cadmium.[28] Ferrarello and coworkers used SEC coupled to a quadrupole or a double-focusing sector field ICP-MS for the speciation of MLPs of the mussel *Mytilus edulis.*[29] The sector field instrument allowed the determination of MLPs in natural populations, a task that has been rarely accomplished due to the comparatively low basal level of these proteins and their complexed metals. Furthermore, the multielement capability of ICP-MS was applied for the simultaneous speciation of Cu and Zn in the corresponding fraction of Cd (Cd is the strongest inductor of MLP synthesis).

The binding of several elements in the tissues of largemouth bass (*Micropterus salmoides)* was investigated by Jackson *et al.*[30] Specific elements investigated by SEC-ICP-MS included Se, As, Cu, Cd and Zn as the employed instrumentation allowed data on multiple trace elements to be collected simultaneously. This analytical scheme hence allowed the elucidation of any interactions between trace elements within the various protein fractions and results showed Cu, Zn and Cd were bound to metallothionein in the liver, gill and, to a lesser extent, the gonad tissue. Furthermore, the existence of selenoproteins was investigated and the major fraction of selenium was found to be associated with high molecular weight species in the gonad tissue. Se is an essential element for animals and a number of proteins, such as glutathione peroxidase and formate dehydrogenase, utilize selenocysteine in the active site. Selenomethionine seems to be incorporated unspecifically in proteins at the methionine position and consequently, a large number of proteins may contain selenium. Onning and Bergdahl also used SEC-ICP-MS to study soluble selenium compounds in fish[31,32] with the goal of determination of the distribution of low and high molecular weight selenium compounds and subsequent elucidation of bioavailability.

Nischwitz and coworkers studied trace element speciation of porcine liver samples by SEC-ICP-MS.[33] Interestingly, they employed subcellular fractionation of the liver homogenate to provide additional information on the location of the extracted species. The fractions were defined by the centrifugation speed and so by differential centrifugation, it was possible to prepare at least three fractions from the liver homogenate which corresponded to cellular components: a cytosolic fraction, a microsomal fraction and a mitochondrial/lysosomal fraction. Each fraction was subsequently analyzed by SEC-ICP-MS.

Suzuki *et al.*[34] used SEC-ICP-MS for the speciation of Se-containing biological constituents. It was determined that diets with different Se contents induced changes in the distributions of Se-containing constituents more so in urine, kidney and liver samples than in plasma and red blood cell samples. In other work performed by this same group, selenoprotein P was specifically examined and in particular, its interaction with transition metal ions such as Ag, Cd and Hg.[35] This interaction between transition metals and selenium in the bloodstream was studied in vitro by means of SEC-ICP-MS using the selenoprotein P fraction prepared from the serum of rats. Results indicated that first, on exposure to such, the transition metal and selenide formed a unit complex, which then in turn was bound to selenoprotein P. Lack of retention time shift of the protein translated to a lack of transition metal complex binding. Further work used SEC-ICP-MS to reveal the mechanism underlying the specific binding of a Hg-selenide complex on selenoprotein P.[36] Principle binding sites were examined in vitro by use of a competitive binding assay.

Artelt and coworkers studied the bioavailability of the fine dispersed elemental platinum emitted by automobile exhaust catalytic converters.[37] Synthetic particles (aluminum oxide deposited with platinum) were applied to laboratory animals in two doses by intratracheal instillation. Of the platinum determined to be bioavailable, SEC-ICP-MS results demonstrated approximately 90% was bound to macromolecules, such as proteins, and approximately 10% was present as low molecular weight species, most likely in the form of Pt II and/or Pt IV complexes.

Microorganisms

Leopold and Fricke analyzed the trace elements in the active center of previously uncharacterized membrane proteases of bacterial origin.[38] Two different membrane proteases from *Bacillus cereus* and *Pseudomonas aeruginosa* were characterized to be zinc metalloproteases using SEC-ICP-MS. Interestingly, they found that the presence of nonionic detergents (as used to maintain the solubility and activity of the proteins in solution during purification or as components of the eluent) can influence the distribution of trace elements during the SEC separation. Therefore, they recommended that the use of these substances should be avoided during enzyme purification for metal analyses or they should be exchanged later with zwitterionic and ionic detergents with stronger dissociating properties.

SEC-ICP-MS was used in the analysis of extracts of selenized yeast by Casiot and coworkers.[39] The ultimate goal of the work was the determination of the most efficient extraction procedure without modification of the chemical form of the selenium species nor disturbance to the macro-equilibrium between the various species present. The different extraction procedures yielded different classes of selenium species and hence SEC proved to be an advantageous technique for screening the selenium species in the yeast extracts since it combines the satisfactory resolution of the small water-soluble selenium species with the possibility of monitoring the high molecular mass fraction. The separation of the small compounds with similar masses was attributed to secondary adsorption and ion-exchange effects.

Vacchina *et al.*[40] used SEC-ICP-MS for screening the changes in the bioligand composition of wood-rotting fungi as a function of their exposure to copper. Wood-rotting fungi are much studied in the field of environmental biotechnology due to their ability to degrade organic contaminants *in situ*, however, this capacity is negatively affected by the presence of heavy metals, such as copper that block the lignin-degrading enzymes present in the fungi. Results of four species studied generally indicated no bioinduction of a cysteine rich ligand, but rather a passive resistance mechanism, e.g. adsorption of Cu on the cell walls or complexation by preexisting ligands.

Humans

The toxic activity of some trace metals such as Cd and Pb is thought to relate to their ability to compete with essential elements in proteins or to bind to DNA. Wang and coworkers used a magnetic sector ICP-MS coupled to the effluent of an SEC column in pursuit of identification of elements in the proteins of human serum as well as in DNA fragments.[41] The mass spectrometer allowed observation of elements at ambient levels in human serum in addition to providing the spectral resolution necessary to measure Cr and Fe at their major isotopes. Results showed that monitoring Cd, Cu and Zn in serum did not result in peaks in the appropriate molecular weight range for metallothioneins (~10 kDa) which promotes speculation that these elements are not stored in metallothioneins in serum or that the metallothioneins are bound to other larger proteins. Additionally, results indicated complete binding of Pb, Cd, Mn and Fe to DNA fragments as well as binding of Cr (VI) after reduction to Cr (III)/oxidation of DNA.

Richarz *et al.*[42] used SEC-ICP-MS in the investigation of the speciation of trace elements in the brains of patients with Alzheimer's disease. Post-mortem samples from Alzheimer's disease brains and from brains of a control group were investigated to elucidate changes in the trace element distribution during the pathological process. Special emphasis was placed on metallothioneins and a comparison between Alzheimer's disease and control brains showed a significant difference concerning several metallothionein metal levels, leading to the assumption that oxidative processes occurred in Alzheimer's brain samples.

In a later work, the same authors[43] investigated protein-bound trace elements in human cell cytosols (soluble proteins) of different organs as well as the same type of organ in different pathological states. This work was based on the assumption that the cells of different organs contain different proteins to perform their respective specialized functions; final identification of the proteins was performed by means of specific protein assays with collected fractions from the SEC column. Different elemental profiles were obtained for different organs of the same patient and for the same organ of patients with different diseases. Therefore, the authors concluded that metalloproteins and their bound metals may be used as biological markers for physiological differences or pathological changes in human tissue. Interestingly, they also found that high iron content may be linked to high hemoglobin content and subsequent high blood supply within specific organs.

Pb-bound ligands in human amniotic fluid were studied by SEC-ICP-MS in work performed by Hall et al.[44] This work was done to support a study that examined maternal-fetal transfer of Pb by comparing Pb-bound ligands in amniotic fluid with those in maternal plasma. Using protein standards, Pb-containing ligand identifications included ceruloplasmin, pre-albumin and a Zn-peptide; multielement detection was employed to verify other known elements in each of the proteins.

SEC-ICP-MS has also been applied for the investigation of drug-protein interactions and binding; such studies are critical for elucidating mechanisms of drug action, toxicity and metabolism. For example, patients undergoing cisplatin chemotherapeutic treatment may experience a decrease of hemoglobin in the blood and Mandal et al. therefore used this technique to study the interaction of cisplatin and hemoglobin.[45] SEC was used to separate free and protein-bound cisplatin and ICP to monitor simultaneously ^{195}Pt and ^{57}Fe; results demonstrated the presence of hemoglobin bound platinum complexes which may explain the mechanism for the hemoglobin reduction. In a similar vein, Szpunar and coworkers studied the interactions of several Pt and Ru-based drugs with serum proteins.[46]

SEC coupled to ICP-MS has also found many important applications in the investigation of infant nutrition. Bratter and coworkers used SEC-ICP-MS to study the relationship between breast milk and maternal dietary intake as well as to investigate the binding pattern of trace elements in formulas as compared to breast milk.[47] They discovered that this binding pattern was significantly different and depended on the main component (cow's milk or soy), its processing (hydrolysis) and the chemical form of the added compounds. Additionally, they noted that breast milk samples from different regions of the world showed comparable shapes for the elution profiles and for Mo and Se, a dependence on the regional maternal dietary intake. Coni *et al.* studied a more extensive list of elements with regard to the protein fraction with which they were bound[48] and in later work by Martino and coworkers, UV detection was additionally utilized for protein identification.[49]

Fernandez and Szpunar specifically investigated iodine species in various types of milk (human as well as cow, goat and infant formulas) by SEC-ICP-MS.[50] Iodine is added to the cow feed in order to enhance production of milk and meat and this makes milk an important source of iodine. As milk cannot be analyzed directly by SEC because it is an emulsion containing solid particles, work was directed toward the development of a sample preparation procedure that would allow the quantitative transfer of iodine-containing species. An approach based on the extraction of iodine-containing species into an aqueous phase (whey) that could be separated from the solid particles (caseine) and fat by ultracentrifugations was investigated. Results showed that a second species in addition to iodide existed in human milk and infant formula and may represent iodine attached to a macromolecular compound by coordination bonds.

Conclusion and Future Perspectives

Size-exclusion chromatography, when combined to inductively coupled plasma mass spectrometry as a detection system, becomes one of the most powerful analytical tools to study elemental distribution in natural and synthetic macromolecules. Many recently growing interdisciplinary research interests of wide diversity can benefit from such a simple and convenient technique. For example, identification and characterization of metalloproteins and their mechanistic pathways in biological system present a real challenge to the analytical chemists as well as the researchers in related fields. The nature and the speciation of metals play a crucial role, particularly in metallo-enzymes as well as in the complex biocatalytic processes that form the basis for developing resistance to antibiotics in bacteria as well as for the existence of the higher organisms. Moreover, many metalloproteins are labile with respect to the coordinately bonded metal ion complexation and requires very gentle extraction and separation conditions to preserve the speciation status. Due to minimum contribution of enthalpy from physico-chemical interaction between analyte and stationary phase, size-exclusion chromatography is the most suitable separation technique for such labile species. Therefore, it opens up the possibility of investigating novel metalloproteins presently unidentified due to the low detention capability and other limitations of the traditional proteomic approaches. However, size-exclusion chromatography does not have great enough resolution capability to separate such a complex biological mixture. Other orthogonal separation techniques are subsequently required to achieve the separation down to the individual component level. Moreover, ICP-MS can not provide any structure specific information. In fact, a multidimensional hyphenation approach with several orthogonal separations and detection techniques is necessary to cope with such a challenge. Nonetheless, SEC-ICP-MS can provide very important preliminary information as the basis for initiating such an elaborate multidimensional approach.

References

(1) Ellis, L. A.; Roberts, D. J. *Journal of Chromatography A* **1997**, *774*, 3-19.

(2) Wang, J.; Dreessen, D.; Wiederi, D. R.; Houk, R. S. *Analytical Biochemistry* **2001**, *288*, 89-96.

(3) Szpunar, J. *Analysts* **2000**, *125*, 963.

(4) Regnier, F. E. *Science* **1983**, *222*, 245.

(5) Mota, A. M.; Simaes Goncalves, M. L. *Elemental Speciation in Bioorganic Chemistry*; John Wiley and Sons: New York, 1996.

(6) Lobinski, R.; Chassaigne, H.; Szpunar, J. *Talanta* **1998**, *46*, 271-289.

(7) Szpunar, J. *Trends in Analytical Chemistry* **2000**, *19*, 127-137.

(8) Szpunar, J.; Lobinski, R. *Analytical and Bioanalytical Chemistry* **2002**, *373*, 404-411.

(9) Prange, A.; Schaumloffel, D. *Analytical and Bioanalytical Chemistry* **2002**, *373*, 441-453.

(10) McSheehy, S.; Mester, Z. *Trends in Analytical Chemistry* **2003**, *22*, 311-326.

(11) Schmitt, D.; Muller, M. B.; Frimmel, F. H. *Acta Hydrochimica and Hydrobiologica* **2000**, *28*, 400-410.

(12) Haraguchi, H.; Itoh, A.; Kimata, C.; Miwa, H. *The Analyst* **1998**, *123*, 773-778.

(13) Hils, A.; Grote, M.; Janssen, E.; Eichhorn, J. *Analytical and Bioanalytical Chemistry* **1999**, *364*, 457-461.

(14) Sadi, B. B. M.; Wrobel, K.; Wrobel, K.; Kannamkumarath, S. S.; Castillo, J. R.; Caruso, J. A. *Journal of Environmental Monitoring* **2002**, *4*, 1010-1016.

(15) Wrobel, K.; Sadi, B. B. M.; Wrobel, K.; Castillo, J. R.; Caruso, J. A. *Analytical Chemistry* **2003**, *75*, 761-767.

(16) Oedegard, K. E.; Lund, W. *Journal of Analytical and Atomic Spectroscopy* **1997**, *12*, 403-408.

(17) Szpunar, J.; Pellerin, P.; Makarov, A.; Doco, T.; Williams, P.; Medina, B.; Lobinski, R. *Journal of Analytical and Atomic Spectroscopy* **1998**, *13*, 749-754.

182

(18) Szpunar, J.; Pellerin, P.; Makarov, A.; Doco, T.; Williams, P.; Lobinski, R. *Journal of Analytical and Atomic Spectroscopy* **1999**, *14*, 639-644.

(19) Lecchi, P.; Gupte, A. R.; Perez, R. E.; Stockert, L. V.; Abramson, F. P. *Journal of Biochemical and Biophysical Methods* **2003**, *56*, 141-152.

(20) Vacchina, V.; Polec, K.; Szpunar, J. *Journal of Analytical and Atomic Spectroscopy* **1999**, *14*, 1557-1566.

(21) Leopold, I.; Gunther, D. *Analytical and Bioanalytical Chemistry* **1997**, *359*, 364-370.

(22) Klueppel, D.; Jakubowski, N.; Messerschmidt, J.; Stuewer, D.; Klockow, D. *Journal of Analytical and Atomic Spectroscopy* **1998**, *13*, 255-262.

(23) Koplik, R.; Pavelkova, H.; Cincibuchova, J.; Mestek, O.; Kvasnicka, F.; Suchanek, M. *Journal of Chromatography B* **2002**, *770*, 261-273.

(24) Mestek, O.; Kominkova, J.; Koplik, R.; Borkova, M.; Suchanek, M. *Talanta* **2002**, *57*, 1133-1142.

(25) Kannamkumarath, S. S.; Wrobel, K.; Wrobel, K.; Vonderheide, A. P.; Caruso, J. A. *Analytical and Bioanalytical Chemistry* **2002**, *373*, 454-460.

(26) Mounicou, S.; Szpunar, J.; Lobinski, R.; Andrey, D.; Blake, C. J. *Journal of Analytical and Atomic Spectroscopy* **2002**, *17*, 880-886.

(27) Ferrarello, C. N.; Campa, M. R. F. d. l.; Sanz-Medel, A. *Analytical and Bioanalytical Chemistry* **2002**, *373*, 412-421.

(28) Tessier, C.; Blais, J. S. *Ecotoxicology and Environmental Safety* **1996**, *33*, 246-252.

(29) Ferrarello, C. N.; Campa, M. R. F. d. l.; Carrasco, J. F.; Sanz-Medel, A. *Analytical Chemistry* **2000**, *72*, 5874-5880.

(30) Jackson, B. P.; Allen, P. L. S.; Hopkins, W. A.; Bertsch, P. M. *Analytical and Bioanalytical Chemistry* **2002**, *374*, 203-211.

(31) Onning, G.; Bergdahl, I. A. *The Analyst* **1999**, *124*, 1435-1438.

(32) Onning, G. *Food Chemistry* **2000**, *68*, 133-139.

(33) Nischwitz, V.; Michalke, B.; Kettrup, A. *Journal of Analytical and Atomic Spectroscopy* **2003**, *18*, 444-451.

(34) Suzuki, K.; Itoh, M.; Ohmichi, M. *Journal of Chromatography B* **1995**, *666*, 13-19.

(35) Sasakura, C.; Suzuki, K. T. *Journal of Inorganic Biochemistry* **1998**, *71*, 159-162.

(36) Suzuki, K. T.; Sasakura, C.; Yoneda, S. *Biochimica and Biophysica Acta* **1998**, *1429*, 102-112.

(37) Artelt, S.; Creutzenberg, O.; Kock, H.; Levsen, K.; Nachtigall, D.; Heinrich, U.; Ruhle, T.; Schlogl, R. *The Science of the Total Environment* **1999**, *228*, 219-242.

(38) Leopold, I.; Fricke, B. *Analytical Biochemistry* **1997**, *252*, 277-285.

(39) Casiot, C.; Szpunar, J.; Lobinski, R.; Potin-Gautier, M. *Journal of Analytical and Atomic Spectroscopy* **1999**, *14*, 645-650.

(40) Vacchina, V.; Baldrian, P.; Gabriel, J.; Szpunar, J. *Analytical and Bioanalytical Chemistry* **2002**, *372*, 453-456.

(41) Wang, J.; Houk, R. S.; Dreessen, D.; Wiederin, D. R. *Journal of the American Chemical Society* **1998**, *120*, 5793-5799.

(42) Richarz, A. N.; Bratter, P. *Analytical and Bioanalytical Chemistry* **2002**, *372*, 412-417.

(43) Richarz, A. N.; Wolf, C.; Bratter, P. *The Analyst* **2003**, *128*, 640-645.

(44) Hall, G. S.; Zhu, X.; Martin, E. G. *Analytical Communications* **1999**, *36*, 93-95.

(45) Mandal, R.; Teixeira, C.; Li, X. F. *The Analyst* **2003**, *128*, 629-634.

(46) Szpunar, J.; Makarov, A.; Pieper, T.; Keppler, B. K.; Lobinski, R. *Analytica Chimica Acta* **1999**, *387*, 135-144.

(47) Bratter, P.; Blasco, I. N.; Bratter, V. E. N. d.; Raab, A. *The Analyst* **1998**, *123*, 821-826.

(48) Coni, E.; Bocca, B.; Galoppi, B.; Alimonti, A.; Caroli, S. *Microchemical Journal* **2000**, *67*, 187-194.

(49) Martino, F. A. R.; Sanchez, M. L. F.; Medel, A. S. *Journal of Analytical and Atomic Spectroscopy* **2002**, *17*, 1271-1277.

(50) Sanchez, L. F.; Szpunar, J. *Journal of Analytical and Atomic Spectroscopy* **1999**, *14*, 1697-1702.

Chapter 11

Size-Exclusion Chromatography Combined with Chemical Reaction Interface Mass Spectrometry for the Analysis of Complex Mixtures of Proteins

Paolo Lecchi and Fred P. Abramson

School of Medicine and Health Sciences, George Washington University, Ross Hall, Room 610, 2300 I Street, N.W., Washington, DC 20037

In size-exclusion chromatography (SEC) analytes are separated based on their molecular size. A size-dependent separation procedure, able to fractionate any class of molecules according to a predictable and universal parameter, could be particularly useful for any comprehensive analysis ("-omics") of heterogeneous mixtures. Combining SEC to a universal detection system such as chemical reaction interface mass spectrometry (CRIMS) results in an analytical scheme able to monitor and quantify any isotope-labeled analyte in complex biological mixtures. This report describes an innovative use of SEC as a critical component in multidimensional separation schemes for proteomics.

Introduction

Proteomics (as any "-omics" analysis) is aimed at providing comprehensive characterization of a biological system in its entirety at the molecular level. However, from the analytical chemistry point of view, any component of a proteome (i.e., any single protein) may be considered as a potential analytical problem (i.e., analyte). Taking in account the huge number of analytes present, their heterogeneity, and the extremely broad range of concentrations, the comprehensive analysis of the whole "proteinaceous material" contained in a complex organism may be seen as an unmanageable task. Indeed, it has been argued that none of the analytical methodologies currently in use will ever be able to provide the complete description of the proteome from a complex organism (1). Hence, the development of new methods for proteomics is one of the more active fields of research in analytical biochemistry.

Conventionally, proteomic analyses are performed by combining a powerful separation scheme, such as two-dimensional polyacrylamide gel electrophoresis (2D-PAGE) with mass spectrometry (MS). It is quite remarkable that, while the performance of MS been greatly improved over the last decade, 2D-PAGE, at almost 30 years from its original development (2), still represents the paradigm for high resolution separation of complex protein mixtures. Briefly, in 2D-PAGE a protein extract is resolved on a flat gel of polyacrylamide according two independent parameters, namely the isoelectric point and the molecular size, yielding several hundred individual spots that can be ultimately identified by MS. The capability of 2D-PAGE to resolve mixtures of intact proteins is still unmatched by any other separation technique, and it can be further enhanced by preliminary treatments of the sample (3). Nevertheless, it is well known that 2D-PAGE has several practical limitations, and does not cover the full spectrum of proteins contained in a typical biological system (4). For this reason, alternative approaches have been introduced in which the gel-based two-dimensional separation is replaced by the combination of two or more liquid chromatographic steps (5).

Similarly to 2D-PAGE, multi-dimensional chromatographic schemes are aimed at resolving complex mixtures of proteins into their individual components. Size-Exclusion Chromatography (SEC), with its low resolving power, may seem unsuitable for conventional chromatographic schemes for proteomics. Nevertheless, analytical strategies where SEC is one of the techniques used in a multidimensional scheme for proteomics have been reported. As an example: Opiteck et al. have connected up to twelve SEC columns, followed by reversed phase chromatography, to obtain dispersion power enough to fractionate the entire proteome of *E. coli* (6). In a more specific application, SEC was combined to capillary isoelectrofocusing (cIEF) in a non-denaturing multidimensional separation procedure, with the potential to separate native protein complexes (7).

The challenge of proteomics clearly is not restricted to just implementing the separation scheme. In fact, even when the separation processes achieve superior resolving power, the characterization of each and every single component in each and every fraction may still be considered impractical. A more effective and manageable analytical perspective could be that of looking at the system in its dynamic dimension, e.g. monitoring only the differences occurring in a biological system when it is represented in two different conditions. Several schemes for such a "differential proteomics" approach have been described and recently reviewed (8, 9). Most utilize a common strategy consisting of labeling a biological system, or a protein extract, with two different isotopic forms of the same element (e.g., ^{13}C and ^{12}C) one for each of the given physiological or physio-pathological conditions under evaluation. For simplicity, we will call the sample labeled with the naturally most abundant isotope (e.g. ^{12}C) "unlabeled", and the sample labeled with the alternative isotopic form (e.g. ^{13}C) "labeled". The two protein extracts are then combined to generate a single sample, in which each protein is represented in both the labeled and the unlabeled form. The mass spectrometric analysis of this sample will result in spectra containing a pair of peaks for each analyte, (i.e., labeled and unlabeled) with the difference in molecular weight (MW) between the two peaks being directly related to the number of isotopic tags introduced in the analyte. Moreover, the relative intensity between the two peaks indicates the relative amounts of the analyte in the labeled and the unlabeled samples, i.e., in the two conditions of the biological system under evaluation. Hence, evaluating the peak pairs and their relative intensity allows identifying those species that are differentially expressed in the two conditions considered, within a background of unmodified proteins.

An alternative way to perform a differential analysis is to measure the isotopic tags directly, rather than the MW shifts that they generate on each labeled analyte. This kind of analysis can be performed by using a "universal" detection scheme able to specifically monitor isotopic tags without being affected by the chemical nature of the analyte.

Over the past years we developed a universal detection scheme that relies on chemical reaction interface mass spectrometry (CRIMS) (10). As illustrated in Figure 1, in CRIMS the complexity of the sample is not an obstacle to the analysis because the analytes do not enter the mass spectrometer intact, but only after decomposition via chemical reaction. This means that all the organic molecules, regardless of their nature, size, or concentration, are transformed into a predictable set of low MW compounds. CRIMS does not provide any information about the molecular weight of the intact species. However, the decomposition strategy provides a different set of parameters for the characterization of the analyte. For example CRIMS can detect the presence of specific isotopic tags such as: ^{2}H, ^{13}C, ^{15}N, ^{18}O, ^{34}S; or infrequent elements such as: S, P, Cl, Br . Also, CRIMS is an efficient way to monitor isotopic ratios such as: ^{2}H/^{1}H, ^{13}C/^{12}C ^{15}N/^{14}N, ^{18}O/^{16}O, ^{34}S/^{32}S.

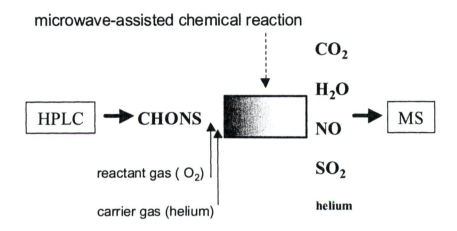

microwave-assisted chemical reaction

HPLC → CHONS

reactant gas (O_2)

carrier gas (helium)

CO_2

H_2O

NO

SO_2

helium

MS

Figure 1: Scheme of a CRIMS apparatus. A chemical reaction interface posted between the liquid chromatography and the mass spectrometric detector continuously transforms any bio-organic analyte (composed of common organic element, e.g. C, H, N, O, S) in a predictable set of low molecular weight compounds that are analyzed by mass spectrometry.

We previously showed that SEC and CRIMS provide a valuable combination able to separate different classes of molecules and to detect species containing specific isotopic or elemental labels (11,12). Increasing the complexity of the sample up to that of whole protein extracts leads to a need for higher chromatographic resolution and so the coupling of multiple separation techniques may be required. The main objective of the experiments described in this manuscript is the implementation of multidimensional chromatographic schemes for proteomics, suitable for the CRIMS detector.

Experimental Section

To test the analytical scheme proposed we utilized a whole protein extract from *E. coli* (strain BL21). Briefly, 200 ml of bacterial culture were grown overnight in M9 minimal essential medium containing glucose (Sigma, St. Louis, MO) or $^{13}C_6$-glucose (Spectra Stable Isotopes, Columbia, MD) as the only source of carbon. In both cases, bacteria were pelleted by centrifugation (10,000 g for 10 minutes), resuspended in 5 ml of water, and disrupted by sonication at 4 °C (30 seconds, 3 times). After the addition of Tris-HCl, pH 7.0, to a final concentration of 20 mM, lysates were equilibrated for four hours at 4 °C, under gentle rocking. Each lysate was then centrifuged (15,000 g for 30 min) and the supernatant was filtered through a 0.22-micron filter and finally dialyzed overnight against water, using a 7 kDa cut-off dialysis cassette (Pierce, Rockford, IL). Five ml of a solution containing approximately 2 mg/ml of proteins were obtained from each bacterial culture.

The first scheme that was evaluated is the pairing of liquid isoelectrofocusing (IEF) with chromatography. Bacterial protein extracts were fractionated first by liquid IEF using a Rotofor-cell (Bio-Rad, Hercules, CA). Four ml of the "unlabeled" bacterial extract were loaded in the electrophoretic cell and fractionated at constant power (12 W) for three hours in 50 ml of a solution containing 2% ampholyte, pH 3-10 (Bio-Rad). Twenty fractions were collected and stored at 4 °C for further separation. Aliquots of sixteen IEF fractions (fractions number three through eighteen) were analyzed by SEC, using a TSK G3000SWXL 7 x 300 mm column (TosoHaas, Montgomeryville, PA). The mobile phase for SEC was KH_2PO_4 50 mM, NaCl 200 mM, pH 6.5/methanol in a 9:1 ratio. For comparison, aliquots obtained from the same sixteen fractions were also analyzed by reversed phase chromatography, using a Vydac C18 TP54 150x4.6 mm column and a linear gradient of water/acetonitrile (both containing 0.1% TFA) from 30% to 70% over 60 minutes. Both separations were monitored by a UV detector set at 280 nm. Results obtained from theses two multidimensional separations were plotted in a contour graph to resemble a 2D-PAGE separation. The program used for graphical reconstruction of data was Slide Write Plus (Advanced Graphic Technologies, Encinitas, CA).

In a second set of experiments, SEC was employed as the first chromatographic dimension to fractionate a ^{13}C-labeled bacterial protein extract. The separation was performed by connecting two G3000SWXL 7 x 300 mm, with mobile phase 200 mM CH_3COONH_4 in water/acetonitrile 87:13, and a flow rate of 0.2 ml/min. The separation was monitored by UV detection at 280 nm, and fractions were automatically collected every 30 seconds. Aliquots from selected fractions were mixed with a constant amount of a whole protein extract from *E. coli* cultured in unlabeled medium. The resulting samples (i.e., unlabeled extracts spiked with labeled fractions) were then analyzed by HPLC-CRIMS using a weak anion exchange column (TSK-DEAE NPR 4.6 x 35 mm, Toso Haas, Montgomeryville, PA) and a linear 40 minute gradient, from 0.02 to 2 M CH_3COONH_4 containing 10% acetonitrile. CRIMS detection was optimized to detect ^{12}C and ^{13}C (as $^{12}CO_2$ at m/z 44 and $^{13}CO_2$ at m/z 45, respectively). Analytes carrying isotopic tags were monitored in the enrichment trace, obtained by subtracting the ^{13}C natural abundance (i.e. 1.19% of ^{12}C) from the m/z 45. Hence, only species containing an amount of ^{13}C exceeding the natural abundance are visualized in the enrichment trace.

Discussion

2D-PAGE separates protein mixtures according to two independent molecular properties, molecular size and isoelectric point. Replacing the gel-based separation with a "free-flowing" separation approach would certainly overcome some of the drawbacks intrinsically associated with the use of flat gels. In our scheme, to enrich for relatively low abundance species we combined isoelectrofocusing (IEF) to SEC. This combination separates a protein extract according to the same molecular properties that are at the base of the separation by 2D-PAGE. Also, both IEF and SEC are non-denaturing procedures; therefore, they may be suitable for the separation of intact protein complexes (7).

Despite promising results, it is quite apparent that the combination of IEF and SEC clearly does not provide the same level of resolution that is typically obtained with 2D-PAGE. A graphical representation of this finding is shown in Figure 2, which is a contour map obtained by combining the SEC analyses of each of 16 fractions obtained from IEF. Although several adjustments to the procedure could be devised to improve the separation scheme (e.g., performing IEF in a narrower pH range, or using finely tuned SEC conditions) as far as resolution is concerned this scheme does not represent a viable alternative to 2D-PAGE. A more effective resolving power is achieved when IEF is followed by reversed phase chromatography, as shown in Figure 3 which is a contour map obtained by reversed phase analysis of the same IEF fractions. Similar conclusions have been previously reported (14, 15).

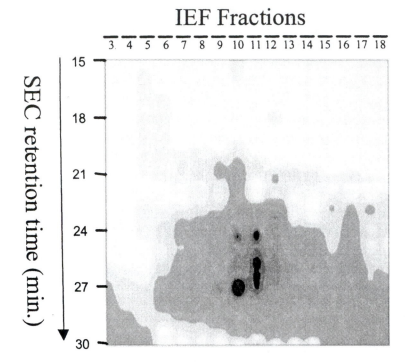

IEF Fractions

Figure 2: IEF-SEC two-dimensional separation of E. coli extract. Each one of the IEF fractions (x-axis, fractions 3 to 18) was analyzed by SEC (y-axis, retention time from 18 to 30 minutes). The z-axis represents the normalized intensity of the chromatographic peaks.

SEC is a universal separation scheme with fundamentally good chromatographic properties but inefficient resolving power; thus, an additional "selection" component will be necessary in order to improve the performance of the separation scheme. Such additional selectivity can be achieved through improved separation, e.g. by combining SEC to an appropriate additional separation technique. Alternatively, the needed further selectivity could be introduced into the scheme by differentially labeling the analytes with isotopic tags. To test the latter analytical strategy, we evaluated the performance of a two-dimensional separation (i.e., SEC-ion exchange) followed by a CRIMS detector in the analysis of a ^{13}C-labeled labeled mixture of bacterial proteins spiked with an analogous mixture of unlabeled bacterial proteins.

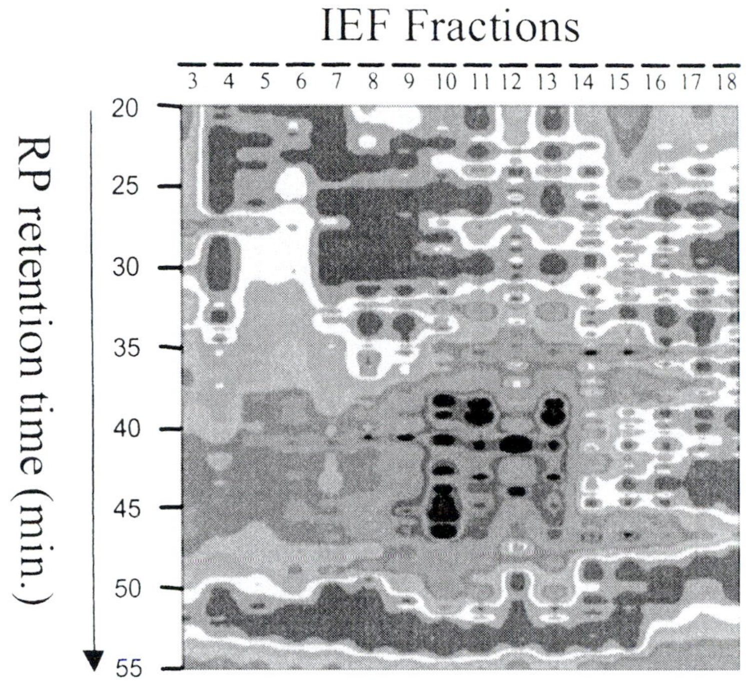

IEF Fractions

Figure 3: IEF-RP two-dimensional separation of E. coli extract. The same IEF fractions were analyzed by reversed-phase chromatography (y-axis, retention time from 25 to 55 minutes).

A whole protein extract of *E. Coli* grown in ^{13}C-enriched medium was first fractionated by SEC, using two serially connected TSK G3000SWXL columns. Fractions were collected automatically every 30 seconds. The chromatogram obtained from the analysis (Figure 4) shows size-dependent separation of the crude bacterial extract. Further dispersion of the biological mixture could be achieved by processing each fraction with a second chromatographic technique based on a separation criterion other than molecular size. The additional separation would enhance the dispersion of the protein mixture (i.e., fewer proteins in each fraction); yet, it also will generate a thousand chromatographic fractions, hardly manageable. As mentioned before, our goal is the implementation of an analytical strategy able to simplify the analysis by sorting out only specific fractions, i.e. those containing the analytes of interest, carrying the isotopic tags.

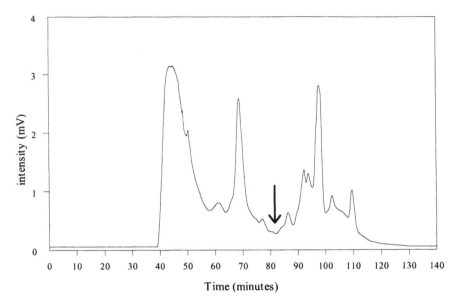

Figure 4: *SEC analysis of a whole protein extract of E. Coli growing in* [13]*C enriched medium. The arrow indicates the fraction that was processed further to yield the results shown in Figure 5. (Reproduced from reference 14. Copyright 2003 Elsevier).*

As illustrated in Figure 1, organic molecules are detected by CRIMS monitoring the channels that are specific for MW of the reaction products generated from the decomposition of the analyte in a microwave induced helium plasma cavity, in the presence of a specific reactant gas. Using this strategy, labeled analytes are also detected in a specific trace. For instance, every ^{12}C-containing analyte generates CO_2, which will be monitored at m/z 44; in the same way, every ^{13}C-containing analyte will generate [13]CO_2 monitored at m/z 45. The natural ^{13}C/^{12}C isotope ratio is 1.19%. Subtracting 1.19% of the ^{12}C trace (i.e., [12]CO_2 with m/z 44) from the ^{13}C trace (i.e., [13]CO_2 with m/z 45) results in an "enrichment trace" which monitors only those analytes containing an amount of ^{13}C that exceeds the natural isotopic abundance (i.e., those carrying isotopic tags).

To validate our analytical scheme, aliquots from selected SEC fractions were used to spike a constant amount of a whole protein extract from *E. coli* cultured in unlabeled medium. The spiked samples were analyzed by ion exchange chromatography coupled with a CRIMS detector. An example of the results obtained is shown in Figure 5, where a 30-second SEC fraction at retention time of 82 minutes (as indicated by an arrow in Figure 4) was the "spike".

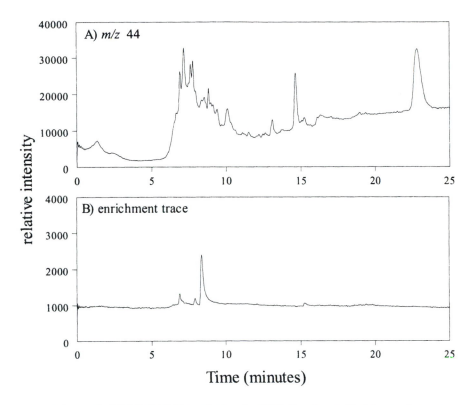

Figure 5. HPLC-CRIMS analysis of a SEC fraction spiked in crude extract obtained from E. coli cultured in unlabeled medium. Panel A: m/z 44 channel (i.e., $^{12}CO_2$). Panel B: enrichment trace obtained by subtraction of 1.19% of m/z 44 from the m/z 45 channel. The enrichment trace monitors only ^{13}C-labeled analytes (i.e., those with ^{13}C contents exceeding the natural abundance).

Figure 5, panel A, shows the "^{12}C trace". This carbon trace monitors all the carbon containing species detected by CRIMS as $^{12}CO_2$ at m/z 44; panel B is the corresponding "enrichment trace", which specifically monitors the labeled analyte. From the chromatographic view point, Figures 4 and 5 show that the combination of SEC and ion exchange has good orthogonality, demonstrated by the fact that a 30-second SEC fraction is dispersed over a 10 minutes span after the second chromatographic separation. Moreover, the high selectivity of the enrichment trace allows detecting stable–isotope labeled analytes even in the presence of an overwhelming amount of interfering material, minimizing the need for chromatographic resolution. Figure 5 shows that in an "enrichment trace" even a small amount of ^{13}C can be easily detected against an overwhelming background of interfering material (i.e., unlabeled proteins detected in the carbon trace).

In addition to monitoring specific isotopes or elements, using an analytical set-up that contains SEC and CRIMS provides direct quantitative information on the amount of the isotope tag detected. In our example we grew bacteria in labeled medium. The direct consequence is that all the biosynthesized molecules were uniformly labeled. When an isotope is always present in the monomeric unit of a polymer, there is a linear relation between the amount of the element in the polymer and the MW of the polymer. Therefore, after uniformly labeling a biopolymer with ^{13}C, the ^{13}C content of each component will correlate well with its MW, so a molar quantitation can be done without requiring extensive standardization procedures. For the analysis of carbon-containing compounds the signal being detected is CO_2, regardless the nature of the analyte. Therefore, any carbon-containing compound with known isotopic abundance, whether enriched or not, can be used to perform the instrument calibration. Even without knowing the identity of the analyte, knowing three parameters would be sufficient to achieve its semi-quantitative determination. The first two parameters (i.e., the area of the $^{13}CO_2$ reaction-product peak and its fractional enrichments) are both obtained by CRIMS analysis. The third parameter (i.e., the MW of the analyte) can be estimated directly by SEC or indirectly (i.e., by an off-line analysis with "conventional" MS). Whenever a separation scheme includes SEC, the information about the MW of the analyte is directly available. Hence, the enrichment trace obtained by CRIMS analysis can be quantitatively interpreted without the need for additional mass spectrometric procedures.

Conclusions

Two multidimensional separation schemes suitable for the analysis of protein mixtures by CRIMS have been evaluated. Despite its limited resolving power, SEC has the advantage of a broad spectrum of applications, and of being able to provide MW information. Moreover, when used in combination with a selective detector such as CRIMS, SEC can be an effective component in the differential analysis of complex protein mixtures. Many proteomics strategies that use stable isotopes can benefit from the analytical strategy described herein.

References

1. Huber, L.A. *Nature Rev. Molec. Cell Biol.* **2003**, 4 , 74-80.
2. O'Farrell, P.H. *J. Biol. Chem.* **1975**, 250, 4007-4021.
3. Gorg, A. et al. *Electrophoresis* **2000**, 21, 1037-1053.
4. Gygi, S.P.; Corthals, G.L.; Zhang, Y.; Rochon, Y.; Aebersold, R. *Proc. Natl. Acad. Sci.* USA **2000**, 97, 9390-9395.
5. Wang, H.; Hanash, S. *J. Chromatogr. B* **2003**, 787, 11-18.
6. Opiteck, G.J.; Ramirez, S.M.; Jorgenson, J.W.; Moseley, M.A. *Anal. Biochem.* **1998**, 258, 349-361.
7. Shen, Y.; Berger, S.Y.; Smith, R.D. *J. Chromatogr A*, 2001, 914, 257-264.
8. Sechi, S.; Oda, Y. *Curr. Opinion Chemical Biol.* **2003**, 7, 70-77.
9. Tao, W.A.; Aebersold, R. *Curr. Opinion Biotech.* **2003**, 14, 110-118.
10. Abramson, F.P. *Mass Spectrom. Rev.* 1994, 13, 341-356.
11. Lecchi, P.; Abramson. F.P. *Anal. Chem.* **1999**, 7, 2951-2955.
12. Lecchi, P.; Abramson, F.P. *J. Chromatogr. A.* **1998**, 828, 509-513.
13. Wall, D.B.; Parus, S.J.; Lubman DM. *J. Chromatogr. B* **2002**, 774, 53-58.
14. Lecchi, P.; Gupte, A.R.; Perez, R.E.; Stockert, L.V.; Abramson F.P. *J. Biochem. Bioph. Method.* **2003**, 56, 141-152.

Acknowledgments

The support from NIH grant RO1 GM 58623 is acknowledged.

We also would like to thank Dr. A.C. Rovescalli for the critical reading of this manuscript.

Chapter 12

Size-Exclusion Chromatography Coupled to Mass Spectrometry and Tandem Mass Spectrometry for Oligomer Analysis

Laszlo Prokai[1], Stanley M. Stevens, Jr.[1], and William J. Simonsick, Jr.[2]

[1]University of Florida, Gainesville, FL 32610–0485
[2]DuPont Marshall R & D Laboratory, 3500 Grays Ferry Avenue, Philadelphia, PA 19146

Coupling size-exclusion (or gel permeation) chromatography to mass spectrometry has been one of the most powerful hyphenated techniques available today for polymer characterization. After reviewing the practice and benefits of the method, we present its extension to involve tandem mass spectrometry (MS/MS) to allow for obtaining detailed structural information on-line with the chromatographic separation of oligomers.

The characterization of polymer structure is important because it provides us with the basics for chemical and physical properties as well as the mechanism of polymerization. The measurement of average properties is no longer adequate to characterize today's polymeric materials. The demand for more extensive polymer analysis and characterization has escalated (1). Hyphenated methods that couple powerful chromatographic separation techniques to information rich detectors such as mass spectrometry (MS) are preferred and necessary for complex polymer analysis. Independently, mass spectrometry can provide the polymer's complete molecular weight distribution, compositional information and end groups, but only for samples that are narrow in their chemical and chain length distributions (2). Higher molecular weight glycidyl methacrylate/butyl methacrylate oligomers were not detected when they coexisted with the lower molecular weight oligomers (3). SEC fractionation was required to characterize the high weight fraction of the acrylic copolymer by electrospray ionization (ESI) mass spectrometry. A size-exclusion chromatograph provides sample fractions that are narrow in both chemical composition and chain length distributions to the mass spectrometer. Hence, microstructure information (chain length, composition and end groups) can be determined for all constituents of a complex polydisperse polymer. This information regarding chain microstructure determines the polymer's bulk physical properties such as rheological behavior or durability/degradation in the environment.

Mass spectrometry plays an important role in the characterization of polymeric materials because of its sensitivity, specificity, and rapid analysis times. Furthermore, soft ionization methods, especially matrix-assisted laser desorption ionization (MALDI) and ESI, in which molecular weight data are preserved, afford predictable data, in contrast to other high-powered analytical techniques, which rely on chromophores, magnetic or dipole moments. ESI with cation doping is capable of ionizing many soluble oxygen-containing polymers. Although tremendous improvements in the number of compounds detected and resolution can be realized through the coupling off-line of SEC with FTMS (3), ESI can also be easily interfaced in an online fashion to condensed phase separation techniques. The soft ionization data provide both the chain length or degree of polymerization and the chemical composition at a given chain length for simple oligomer mixtures (4-6). Unfortunately, ESI will only yield accurate quantitative data when the distributions in chain length and chemical composition are narrow (5, 6) as provided by the SEC. The approach that we have taken to narrow the distributions of a polymer is by on-line coupling to a SEC in which the mechanism of separation is well understood as polymers are predictably separated according to size in solution or their hydrodynamic volume. Furthermore, SEC analysis is the preferred technique to obtain the molecular weight distribution of polymers and oligomers. Ionization has generally been achieved by cation-doping so that the alkali atom adducts are detected. Separation and detection of the fractionated ionized polymer has been accomplished using quadrupoles (7, 8), ion traps (9), time-of-flight analyzers (TOF) (10) and by Fourier-transform mass spectrometry (FTMS) (11). FTMS is the gold standard in providing high resolution and mass accuracy, thus facilitating the direct determination of the ion's charge state, monomer repeat unit and end-group determinations to about $m/z = 10,000$. Furthermore, only 0.5% of the SEC effluent is required for the FTMS.

Figure 1 displays the SEC/FTMS data for poly(methylmethacrylate) (PMMA) 5270 with the differential refractive index (RI) output (Figure 1a) and two representative mass spectra (Figure 1b-c) taken across the RI curve. The mass spectrum seen in Figure 1b was collected at approximately 26.5 minutes, at the beginning of the elution profile. Ions are observed as $[M + nNa]^{n+}$ species

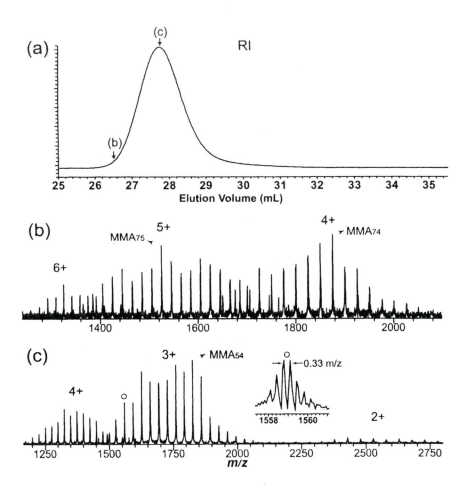

Figure 1. *SEC/RI/ESI/FTMS of PMMA 5270.* **(a)** *RI chromatogram,* **(b)** *ESI mass spectrum at 26.5 min,* **(c)** *ESI mass spectrum at 27.75 min,* **(d)** *log(MW) versus elution volume calibration with selected oligomer profiles (11).*

(d)

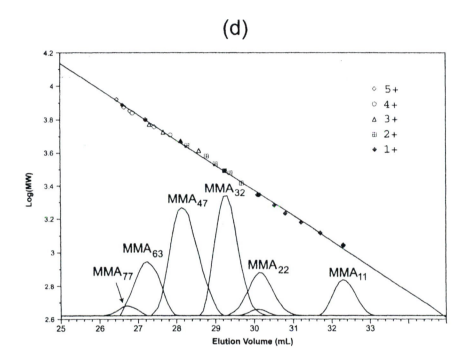

Figure 1. *Continued.*

with overlapping charge state envelopes observed. The oligomers eluting during this 1-s window range in size from 65 to 89 MMA repeat units. Figure 1c was acquired near the apex of the RI response where the largest mass fraction of polymer eluted, at approximately 27.8 minutes. The molecular weights, as expected, are lower and lower charge states are observed. The 43-mer thru 63-mers elute at this time. With a nominal repeat unit mass of 100 Da the charge states are easily determined from the difference in m/z of the oligomers (i.e., 50 m/z for the 2+ charge state). The advantage of the high resolution possible with FTMS is seen in the isotopic resolution shown in the inset of the 46-mer at m/z 1559. The charge state can also be determined directly from the spacing of the isotope clusters. This is important for more complex polymers such as copolymers, where the repeat unit is not so obvious. Striegel et. al. (12) also discussed the direct determination of charge state from the spacings between the ^{13}C isotopes when reporting on the dilute solution behavior of dendrimers. The other advantage of improved resolution is the higher mass accuracy that affords end-group determination. From the exact mass and isotope distribution it is clear that this PMMA has saturated end-groups with little if any vinyl termination (11). The accurate determination of the vinyl group content of an oligomer is important because vinyl-terminated oligomers termed macromonomers are frequently used as building blocks for higher molecular weight architecturally designed polymers (13). The presence of the vinyl group allows for subsequent free-radical polymerization.

Furthermore, a relationship between polymer size and absolute mass from the FTMS data is generated by profiling selected oligomers and computing their retention time from the apex of the peak and plotting against their corresponding molecular weight. Figure 1d profiles the 77-mer, 63-mer, 47-mer, 32-mer, 22-mer and the 11-mer. The relationship between the SEC elution times and the polymer molecular weights is used to generate SEC calibration curves that furnish accurate molecular weights that do not rely upon unrelated calibrants such as linear polystyrenes. Information about polymer architecture (linear, cyclic or branched) can also be furnished (14). For example, dendrimers or highly branched polymers have a more compact structure compared to their linear analogues so that their size versus mass relationship is different.

Figure 2 presents the calibration curves for linear, cyclic and hyperbranched polyesters. The larger molecules of higher molecular weight elute first followed by the smaller molecules possessing lower molecular weights. A plot of the logarithm molecular weight versus the elution time displays a linear relationship. The cyclic materials have a more compact structure than their linear counterparts as we observe that for the same approximate size or elution volume that the cycles are significantly heavier. For the hyperbranched oligomers, dimethylolpropionic acid (DMPA), one observes at low degree-of-polymerization (dp) the elution behavior of the DMPA oligomers resembles that of linear polyesters. However, at higher dp the DMPA oligomers are even more compact than the polyester cycles thus minor size changes as measured by SEC elution time are accompanied by much larger changes in mass.

The hyphenated technique of SEC/MS is also useful for complex mixture analysis. The products formed from the ring opening polymerization of cyclic ethers and esters using single-site metal alkoxide catalysts can produce quite a complex distribution of products that significantly vary in both their dp and end groups (9). SEC/FTMS was needed to characterize these types of complex mixtures. Using SEC/FTMS we were able to identify all the end groups and furnish accurate molecular weights. Furthermore, we used SEC/MS to resolve isomeric propylene glycol oligomers as they differed in their hydrodynamic volume (15).

We have evaluated the use of narrow-bore or micro-SEC (16). This combination offered improvements to the technique, such as low eluent consumption, low cost per column, reduced maintenance requirements, ability to interface to other chromatographic techniques or separation modes and coupling to ESI mass spectrometry without the need for flow splitting. In addition, we observed an increase in chromatographic performance as defined by the resolution factor using microcolumns for SEC compared to conventional size SEC columns (16).

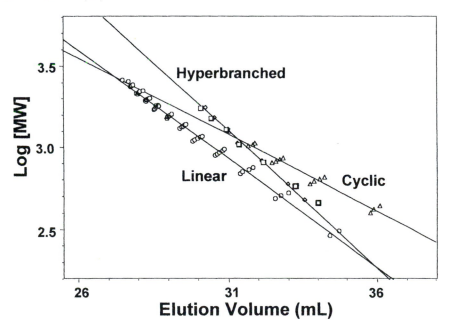

Figure 2. Comparison of calibration curves for linear(circles), cyclic (triangles) and hyperbranched polyesters (squares) obtained by on-line SEC/ESI-MS.

Despite the numerous advantages that we have found using SEC/MS, there are some significant shortcomings of the methodology. Soft ionization mass spectrometry provides little structural details. For example, block versus random copolymers are not distinguished, but yet have vastly different properties. We have furnished structure details about block acrylic copolymers and various polyester polymers (17, 18). Unfortunately, these results were performed using direct ESI, and we were not able to conduct these experiments during the SEC/MS analysis. In the following sections, we report the addition of on-line tandem mass spectrometry to the technique to facilitate the interrogation of structural details of oligomeric species separated by the SEC. Although we will demonstrate the utility of tandem mass spectrometry on SEC-separated materials, other chromatographic techniques that simplify the components entering the mass spectrometer can be used.

Experimental

Chromatographic separation of the oligomeric mixture Triton X-100 was carried out using a two-column set of 500 and 100-Å 30 cm x 7.8 mm i.d Ultrastyragel columns. The mobile phase, tetrahydrofuran (THF) was introduced directly (without splitting) to the mass spectrometer via a Spectroflow 4000 solvent delivery system (Kratos Analytical, Manchester, UK) at a flow rate of 1 mL/min. Triton X-100 samples were dissolved in mobile phase to a final concentration of 0.1 % (w/v) before subsequent SEC–atmospheric-pressure chemical ionization (APCI)/MS and MS/MS analysis. The samples were loaded into a 20 µL loop and injected onto the SEC column through an integrated injection valve controlled by the instrument software of the mass spectrometer.

APCI experiments were performed by interfacing a commercial ion source to a quadrupole ion trap (IT) instrument (LCQ, ThermoFinnigan, San Jose, CA, USA) operated with the Xcalibur (version 1.3) data system software. Oligomer species that eluted off the SEC column were directly infused into the APCI source operating with a vaporizer temperature of 450°C and discharge current of 5 µA. The capillary temperature was maintained at 150°C. The sheath and auxiliary gas flow rates were set to 60 and 10 units, respectively. Full-scan and product-ion mass spectra were obtained using the data-dependent acquisition feature in the Xcalibur software. Full-scan mass spectra were acquired from m/z 300 to 2000 using the automatic gain control mode of ion trapping (target ion count of 5×10^7). Those oligomeric ions that exceeded the threshold level of 1×10^5 counts were then subjected to collision-induced dissociation (CID) using a 2.0 u isolation width and 1.75 V (35% of the maximum value) activation amplitude with helium as the target gas.

Results and Discussion

The application of tandem mass spectrometry for the analysis of complex oligomeric or polymeric systems yields an extremely information-rich, high-throughput technique. In the case of conventional MALDI mass spectrometers, these instruments are generally equipped with TOF analyzers and little can be done to obtain informative tandem mass spectrometric data. Post source decay (PSD), which relies on the inherent metastable decomposition of the precursor ion after the laser desorption/ionization event, can be utilized to furnish structural information, but controlling or manipulating this dissociative process is difficult and time consuming. Recently, hybrid instrumentation (sector-TOF, quadrupole-TOF, or TOF-TOF instruments) has been developed to overcome some of the limitations associated with MALDI-TOF analyzers. It has been shown that some structural information can be obtained from CID spectra of metal-ion/oligomer adducts [poly(methyl methacrylate) and poly(styrene)] generated from a MALDI-sector-TOF (19-21). A majority of MALDI spectra obtained from the analysis of these oligomeric/polymeric systems require the attachment of metal ions, as is the case for mass spectrometric analysis of polymers/oligomers which employ ESI as the ionization method. Given the size of the cation and structure of the oligomer/polymer, CID spectra may be less informative for cation adducts versus their protonated analogues. When the choice of cation dopant is important, one must consider that the binding energy of the cation to the oligomer/polymer decreases with increasing cation size (22) and the predominant signal observed in the CID mass spectra is the loss of the cation from the oligomer/polymer species as the cation size increases. Alkali or metal adduct ions also tend to promote "site-specific" fragmentation that is structurally informative for a particular region of the precursor ion [useful, for example, in C-terminal sequencing of peptides (23)]; however, comprehensive structure determination may be less feasible or impossible, without MS^n capabilites. In any case, even when meaningful CID data are generated, the use of offline SEC fractionation is still necessary for complex mixture analysis and consequently counterproductive to any high-throughput capabilities associated with MALDI instruments. Mass spectrometers equipped to handle conventional atmospheric-pressure ionization (API) interfaces, such as ESI and APCI, are far more favorable in this regard.

Little or no modifications were necessary to allow for interfacing with API sources available on commercial quadrupole, ion-trap and FT-ICR mass spectrometers operating in our laboratories. ESI has been the preferred method of API for polymer analysis. APCI has revealed limitations in terms of mass range and often an excessive fragmentation obscuring molecular ions, although it may produce protonated molecules more beneficial to structural studies via tandem mass spectrometry (MS/MS), as opposed to the cation adducts observed in the ESI mass spectra. [Unlike ESI (7, 8, 11), APCI does not need the

204

introduction of an auxiliary agent and effluent splitting to afford the ionization of oligomers, which simplifies its coupling to SEC.] Figure 3 displays the base-peak chromatogram (BPC) and a full scan mass spectrum at 17.12 min from a SEC/APCI-MS analysis of Triton X-100 obtained on our LCQ ion-trap instrument. The top trace also shows individual selected ion retrievals for the protonated 5-, 10-, 15- and 20-mer of Triton X-100.

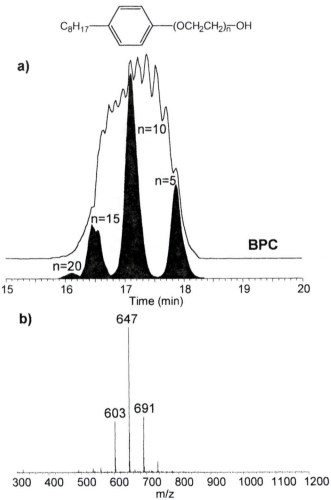

Figure 3. On-line SEC/APCI-MS analysis of Triton X-100: **(a)** Base-peak chromatogram (BPC) and extracted ion chromatograms for the n-mers indicated (m/z 427, 647, 867 and 1087, respectively); **(b)** full-scan mass spectrum collected at 17.12 min.

The novel aspect of this work involves the automated tandem mass spectrometric capability of the LCQ ion trap. With this method, the most abundant ions that elute from the column into the mass spectrometer can be selected for MS/MS analysis on a chromatographic time scale. Assuming an average chromatographic peak width of 20 to 30 sec, multiple (~25-30 on average with conventional 3D ion traps) components that co-elute can be analyzed within that time period given dynamic exclusion is enabled in the acquisition method. Dynamic exclusion prevents multiple selections of the same peak when subsequent full scan mass spectra are acquired by placing an ion that has already been selected for MS/MS on an exclusion list for a user-defined time period. This feature also prevents multiple selections of abundant contaminant peaks that may interfere with mass spectrometric analysis. Based on the oligomer distribution in a single-scan mass spectrum (Figure 3b), the four most intense ions from a full scan mass spectra acquisition were chosen for MS/MS in the data dependent acquisition (DDA) method used for Triton X-100 studies. Figure 4 a) is a CID mass spectrum of the 8-mer at m/z 559 isolated and fragmented in a data dependent fashion from a SEC/APCI-MS analysis. Multiple fragment ion peaks were identified upon analysis of the CID mass

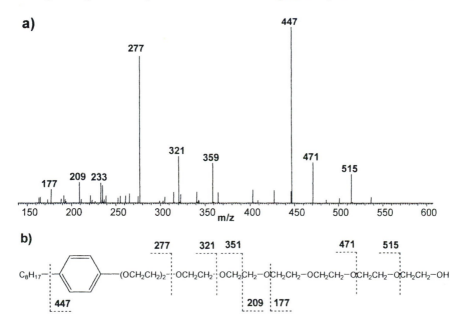

Figure 4. *a) Full scan MS/MS of Triton X-100 oligomer, n=8, (products of m/z 559) recorded via dynamic data acquisition (DDA) following SEC/APCI-MS, and **b**) the possible origin of fragment ions observed.*

spectrum [Figure 4 b)], demonstrating the detailed structural information that can be obtained in a semi-automated, high-throughput manner.

A similar analysis was performed using ESI as the ionization method; however, no protonated species were observed. A cation dopant (Na^+) was required for ion production, and from our data-dependent analysis, MS/MS spectra were much less informative when compared to tandem MS data obtained for the protonated analogues generated from APCI. There were some sodiated fragment ion peaks observed, but >20 MS/MS scans were necessary to produce an appropriate signal. The amount of time required to perform >20 MS/MS scans is not appropriate for data-dependent analysis since the chromatographic elution time for a particular oligomeric species may not dictate such a scan-averaging time period. Moreover, other low-abundance, co-eluting species may never maintain the threshold intensity level necessary to select the ion for MS/MS analysis. Our preliminary studies have, however, indicated that Li^+ attachment affords more efficient fragmentation, which may permit on-line SEC/ESI-MS/MS analyses. Similar findings were reported by Teesch and Adams when studying the collision-induced decompositions of complexes between alkali metal ions and peptides (24). The feasibility of on-line SEC/ESI-MS/MS that relies on Li^+ attachment to generate molecular (precursor) ions from the oligomers is currently under investigation in our laboratory.

Conclusions

The utility of on-line coupling of mass spectrometry with SEC for obtaining information about chemical composition, SEC calibration, and complex mixture analysis has been discussed in this overview. A novel approach that employs MS/MS on-line with SEC using data-dependent acquisition (DDA) on an ion-trap instrument has also been introduced. Results obtained by APCI as a method of ionization for the oligomeric surfactant Triton X-100 have shown the value of the technique.

References

(1) Provder, T.; Barth, H. G.; Urban, M. *Chromatographic Characterization of Polymers: Hyphenated and Multidimensional Techniques*; ACS Advances in Chemistry Series 247; American Chemical Society: Washington, DC, 1995.

(2) Jackson, C.; Simonsick, W. J., Jr. *Curr. Opin. Solid St. M.* **1997**, *2*, 661–667.

(3) Shi, S. D. -H.; Hendrickson, C. L.; Marshall, A. G.; Simonsick, W. J., Jr.; Aaserud, D. J. *Anal. Chem.* **1998**, *70*, 3220–3226.

(4) McEwen, C. N.; Simonsick, W. J., Jr.; Larsen, B. S.; Ute, K.; Hatada, K. *J. Am. Soc. Mass Spectrom.* **1995**, *6*, 906–911.

(5) Prokai, L. *Int. J. Polym. Anal. Charact.* **2001**, *6*, 379–391.

(6) Prokai, L. *in Mass Spectrometry of Polymers*; Montaudo, G.; Lattimer, R. P., Eds.; CRC Press: Boca Raton, FL, 2002; p 149–180.

(7) Prokai, L.; Simonsick, W. J., Jr. *Rapid Commun. Mass Spectrom.* **1993**, *7*, 853–856.

(8) Nielen, M. W. F. *Rapid Commun. Mass Spectrom.* **1996**, *10*, 1652–1660.

(9) Trap-Li, P.; Coleman, D. W.; Spaulding, K. M.; McClennen, W. H.; Stafford, P. R.; Fife, D. J. *J.Chromatogr. A* **2001**, *914*, 147–159.

(10) Liu, X. Michael,; Maziarz, E. Peter; Heiler, David, J.; Grobe, George L *J. Am. Soc. Mass Spectrom.* 2003, 14(3), 195–202.

(11) Aaserud, D. J.; Prokai, L.; Simonsick, W. J., Jr. *Anal. Chem.* **1999**, *71*, 4793–4799.

(12) Striegel, A. M., Plattner, R. D., Willett, J. L., *Anal. Chem.*, **1999**, 71(5), 978-986.

(13) Janowicz, A. H., US Patent, 4,694,054, **1985**.

(14) Simonsick, W. J., Jr.; Aaserud, D. J.; Zhong, W. *Polym. Prepr. (Am. Chem. Soc., Div. Polym. Chem.)*, **2000**, *41*, 657–658.

(15) Antelmann, B.; Chisholm, M. H.; Iyer, S. S.; Huffman, J. C.; Navarro-Llobet, D.; Pagel, M.; Simonsick, W. J., Jr.; Zhong, W. *Macromolecules*, **2001**, *34*, 3159–3175.

(16) Prokai, L.; Aaserud, D. J.; Simonsick, W. J., Jr. *J. Chromatogr. A*, **1999**, *835*, 121–126.

(17) Jackson, A. T.; Scrivens, J. H.; Simonsick, W. J., Jr.; Green, M. R.; Bateman, R. H. *Polym. Prepr. (Am. Chem. Soc., Div. Polym. Chem.)*, **2000**, *41*, 641-642.

(18) Simonsick, W. J., Jr.; Prokai, L. *Proceedings of the 48th ASMS Conference on Mass Spectrometry and Allied Topics* (Long Beach, CA, June 11-15, 2000); American Society for Mass Spectrometry: Santa Fe, NM, 2000; CD-ROM: MODpm.pdf.

(19) Scrivens, J. H.; Jackson, A. T.; Yates, H. T.; Green, M. R.; Critchley, G.; Brown, J.; Bateman, R. H.; Bowers, M. T.; Gidden, J. *Int. J. Mass Spectrom. Ion Process.*, **1997**, *165*, 363-375.

(20) Jackson, A. T.; Yates, H. T.; Scrivens, J. H.; Green, M. R.; Bateman, R. H. *J. Am. Soc. Mass Spectrom.* **1998**, *9*, 269-274.

(21) Jackson, A. T.; Bunn, A.; Hutchings, L. R.; Kiff, F. T.; Richards, R. W.; Williams, J.; Green, M. R.; Bateman, R. H. *Polymer* **2000**, *41*, 7437-7450.

(22) Gidden, J.; Bowers, M. T.; Jackson, A. T.; Scrivens, J. H. *J. Am. Soc. Mass Spectrom.* **2002**, *13*, 499-505.

(23) Lin, T.; Payne, A. H.; Glish, G. L. *J. Am. Soc. Mass Spectrom.* **2001**, *12*, 497-504.

(24) Teesch, L. M.; Adams, J. *Org. Mass Spectrom.* **1992**, *27*, 931-943.

Chemical and Compositional Heterogeneity and Chemical Composition Distribution

Chapter 13

Measuring Compositional Heterogeneity in Polyolefins Using Size-Exclusion Chromatography/ Fourier Transform Infrared Spectroscopy

Paul J. DesLauriers

Chevron Phillips Chemical Company LP, Phillips R&D Center, Highways 60 and 123, Bartlesville, OK 74004

This chapter provides an introductory overview for using size exclusion chromatography (SEC) and on-line, Fourier transform infrared spectroscopy (FTIR) to characterize comonomer content across the molecular weight distribution in polyolefins. The basic spectral aspects of the method, considerations for FTIR as an on-line detector, and the use and limits of on-line SEC-FTIR to detect polymer compositional heterogeneity are addressed. Although on-line SEC-FTIR is a rapid and powerful tool to detect trends resulting from catalyst and process changes, the method's ability to discern a sample's compositional heterogeneity is dependent upon the extent to which the molecular weight distributions of its components overlap. In some cases, SEC-FTIR cannot fully elucidate a sample's compositional heterogeneity. Other techniques such as analytical temperature rising elution fractionation (ATREF) must then be used.

To say that most commercial LLDPE and HDPE polyethylene resins produced today are compositionally heterogeneous is a gross understatement. Primary structural variations in these polymers include chemical structure (end group chemistry and comonomer content), heterogeneity in the molecular weight distributions arising from multi-site catalysts or dual catalyst systems, and topological variations (short and long chain branching architectures). These variations in turn lead to secondary and tertiary structural (morphological) variations that arise during the extrusion and/or manufacturing process of the final product. Of course, the multitude of additive formulations and filler types blended with these polymers further augment the heterogeneities of the final products.

As polyolefins become more complex in their composition, characterization of the polymer's molecular weight and polydispersity by size exclusion chromatography (SEC) using one detector system is no longer sufficient to define the architecture of the resin or the subsequent structure property relationships that shed light on product performance. It is not surprising that considerable effort has been made in recent years to develop analytical techniques that characterize the heterogeneity of polyolefin resins. These techniques include a number of hyphenated methods, many of which are described by other contributors to this text.

In this chapter, the focus is on how comonomer content is characterized across the molecular weight distribution in polyolefins using SEC and on-line Fourier transform infrared spectroscopy (FTIR). Although all of the variations mentioned above play important roles in determining the resin performance properties, molecular weight, comonomer content and its distribution over the polymer's molecular weight are critical to understanding the physical properties of the resin. Molecular weight and its distribution give the polymer its mechanical properties and will influence resin processability. Comonomer content and distribution, at both the inter-molecular and intra-molecular levels, will result in pronounced differences in density as well as in resin performance properties such as thermoxidative stability, stress crack resistance, impact strength, hot tack, heat seal and hexane extractables, to name just a few (1-5).

On-line SEC-FTIR provides a convenient way to gain both MW data as well as comonomer distribution across the MWD. In the following sections, the basic spectral aspects, considerations for FTIR as an on-line detector, and use and limits of this method to detect polymer structural heterogeneity are briefly reviewed. Experimental details for much of the work presented in this chapter can be found in Reference 11.

IR Absorption Bands and Polyolefin Topology

Three absorption regions in polyolefin FTIR spectra are used to glean information about the type and level of short chain branching (SCB) in these resins: the C-H deformation or bending region (1376-1384 cm^{-1}), the methyl and methylene rocking regions (1200 to 800 cm^{-1} and 770 – 720 cm^{-1}, respectively) and the C-H stretching bands found between 2980 and 2950 cm^{-1}. All three-absorption regions and their relative molar absorbtivities are illustrated in Figure 1.

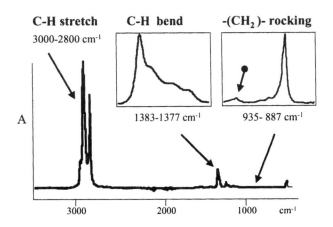

Figure 1. ATR spectrum for an ethylene 1-olefin copolymer film sample showing the three sets of IR absorption bands used to detect methyl (SCB) content.

Historically, peaks arising from the C-H bending modes have been used to determine SCB *(6,7)* in solid-state characterization of polyolefin samples. These bands have medium absorbencies and can be easily detected using either double beam or Fourier transform infrared spectrophotometers and deuterated triglycine sulfate (DTGS) detectors *(7)*. However, both peak position and the measured absorbance for C-H deformations are affected by going to methyl, ethyl, and longer branch lengths *(6)*. The absorption contributions from SCB and chain end methyls, as well as, the change in molar absorptivety, make it difficult to acquire an accurate measurement of the SCB level using this spectral region since the measured value is a weighted average of the various methyl types.

The rocking modes have also been used to determine the type and amount of SCB in polyolefin film samples *(6)*. As in the C-H bending modes, both peak position and molar absorptivity are affected by chain length. However, since the absorbencies of various chain lengths occur in different spectral regions, the use of these vibrational modes has proved useful for identifying specific chain types. For example, Blitz et. al. report *(6)* that various branch types were qualitatively and quantitatively characterized in LLDPE copolymers, LLDPE terpolymers and LDPE resins using these absorption bands. Methyl branches were characterized by an absorbance at 935 cm^{-1}, ethyl branches at 770 cm^{-1}, butyl branches at 893 cm^{-1}, isobutyl branches at 920 cm^{-1}, and hexyl branches at 888 cm^{-1}. Although useful, the molar absorptivity of these bands are very low as illustrated in Figure 1 and typically requires a good spectrometer coupled with a mercury cadmium telluride (MCT) detector in order to properly quantify these peaks.

The C-H stretching bands, (3000 to 2800 cm^{-1}) for simple n-alkanes and polyolefins are the strongest of the three absorptions. This spectral region is composed of a complex over-lapping band system due to vibrations from both methyl and methylene moieties. Four distinct absorption bands are typically observed in simple n-alkanes (8-11). Absorption bands at 2855 and 2928 cm^{-1} are due to symmetrical (CH$_2$ v_s) and asymmetrical (CH$_2$ v_a) stretching of the carbon hydrogen bond in methylene groups, respectively. The other two bands arise from the symmetrical (CH$_3$ v_s) at 2874 cm^{-1} and asymmetrical (CH$_3$ v_a) vibrations at 2957 cm^{-1}. In addition to these four fundamental bands, a possible CH$_2$ combination band (11) is detected as a broad shoulder on the side of the 2928 cm^{-1} methylene peak (~2900 cm^{-1}).

The majority of ethylene 1-olefin copolymers contain isolated short chain branches. That is, the branches off the main chain backbone are not adjacent to each other. When methyl moieties appear as chain ends to isolated short chain branches of four carbons or greater, spectral profiles similar to those found for straight-chain hydrocarbons are observed (i.e., at least five main peaks) (11). However for the branched resins, slight differences in peak positions for the methyl C-H stretch for methyl groups attached to side chains with one methylene unit or less (i.e., methyl and ethyl branches) and a broadening of the 2930 to 2980 cm^{-1} region occur as well. The carbon hydrogen stretching vibration in the methyne moiety is very weak and is reported to occur at ~2890 cm^{-1} (10). Although, this absorption is seldom identifiable due to its low absorptivity and the presence of the CH$_2$ combination bands, it's contribution to the noted broadening effect for short chain branched samples cannot be ruled out.

Lastly, in the case of 1-olefin homopolymer the short chain branches can be considered adjacent to each other. The presence of adjacent SCB branches in particular 1-olefin homopolymers further complicates the FTIR spectra of these samples compared to samples containing only isolated SCB. For example, although the spectrum of poly(1-hexene) is very similar to that observed for n-alkanes and ethylene 1-olefin copolymers, the spectra for polypropylene and , poly(1-butene) both show the presence of more than one absorption peak in the asymmetric methyl stretching region of the spectrum (ca. 2960 cm^{-1}), as well as shifts in other peak absorptions (Figure 2).

Although peak position of the methyl group can vary in some ethylene 1-olefin copolymers and poly (1-olefin) homopolymers, the absoptivities for these moieties are unaffected by chain length (11). This observation and the fact that the C-H stretch in these polymers is the strongest absorption, makes this absorption region ideal for monitoring concentration in on-line SEC-FTIR analysis.

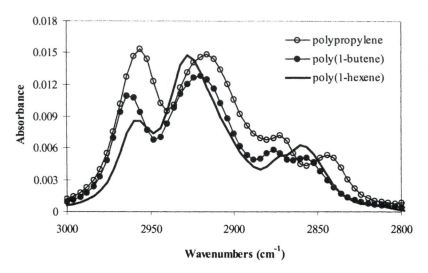

Figure 2. FTIR spectra of poly(1-hexene), polypropylene and , poly(1-butene).
Spectra acquired by SEC-FTIR at the peak max of the eluting sample.

FTIR as an on-line SEC concentration detector

The extent to which the above structural information can be acquired by SEC-FTIR is dependent on the chosen methodology. Two types of SEC-FTIR methods are typically used. In one method, sample eluent from the SEC column is deposited on a rotating germanium disk *(12)*. The solid deposit is subsequently analyzed off-line by FTIR for branching content using absorption bands associated with the bending and/or rocking modes *(13)*. However, deposition heterogeneity, loss of volatile components and other factors can prove problematic *(14)*. In on-line FTIR methods *(11,15-18)*, branching levels in the SEC eluent are measure in a heated flow cell (Figure 3). In this method the absorption characteristics of the solvent dictates which spectral region is accessible for analysis of the polymer itself. Although this latter consideration has been cited as a limiting factor in the use of FTIR detectors with flow cells *(14)*, this methodology remains far more sensitive and flexible than those methods which employ single-beam photometers and detectors fitted with fixed-wavelength interference filters.

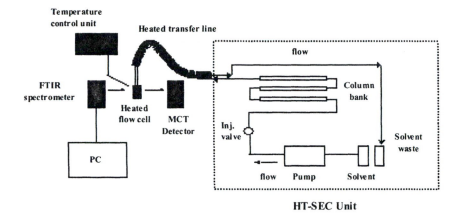

Figure 3. Schematic for a typical SEC-FTIR unit and cell used for on-line SEC-FTIR characterization (11). This is a standard arrangement for high temperature SEC with the obvious addition of the external transfer line, heated flow cell, and narrow band, mercury cadmium telluride (MCT) FTIR detector.

Developments in spectrographic software *(19)* and the availability of a dependable high-temperature flow cell *(20)*, has helped furthered the application of FTIR as an on-line detector for polyolefins. The schematic shown in Figure 4 illustrates a high temperature flow cell well suited for use in this method. Generally speaking, the cell consists of IR transparent windows separated by a spacer. The IR beam passes perpendicular to the effluent flow as the sample enters and exits the heated flow cell. For on-line SEC-FTIR characterization of semi crystalline polyolefins, 1,2,4-trichlorobenzene (TCB) is commonly used as the solvent. However, other high boiling point solvents with the appropriate spectral window and thermal stabilities can be used. Varieties of optical materials are available for use as the IR cell windows under these conditions and include quartz, KBr, ZnSe, and CaF. The cell volumes typically vary between 25 to 70 µL and the optimum path length for high temperature work was found to be 1.6 mm *(21)*. However, a 1 mm path length seems to be adequate for most work *(11,15,16,18)*.

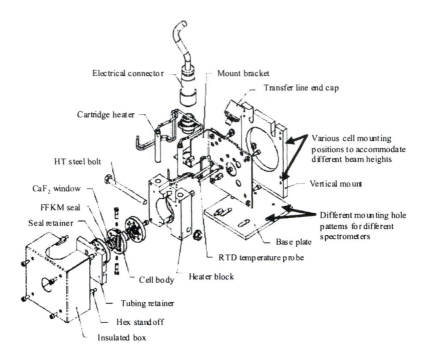

Figure 4. Exploded view of heated flow cell (Polymer Laboratories, Ltd.).

Chromatograms are generated from ratio-recorded transmittance spectra (22). As shown in Figure 5, base line spectra is first acquired over the 3000-2700 cm⁻¹ spectral region. The appropriate absorption band measurement is then acquired and compared against the acquired back-ground. Although 260 background scans (at 8 cm⁻¹ resolution) and up to 50 scans have been used for each MW slice *(18)*, 8 to 16 scans at this resolution for both background and data collection are adequate *(11,16)*. The final chromatogram profile is generated by using the Gram Smitt algorithm or root mean square (RMS) absorption methods. Figure 6 shows a chromatographic profile generated for a polyethylene sample using the RMS method. The level of baseline noise is also shown (insert).

The detection limit for any detector is typically described as when a transducer (detector) signal is twice the noise level of the detector *(23)*. In the case of the FTIR as an on-line SEC concentration detector using the 3000-2700 cm-1 spectral region, the detection limit occurs at 2×10^{-4} RMS absorption units and corresponds to ~3µg/mL polymer sample concentration at the detector.

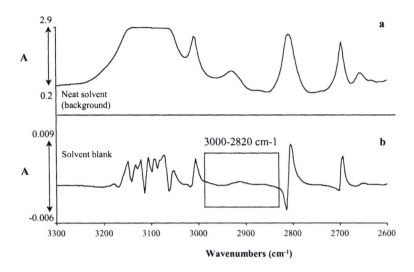

Figure 5. Absorption for the TCB mobile phase over the 3000-2700 cm^{-1} spectral region (a) and the same spectral region after subtracting out the solvent background (b).

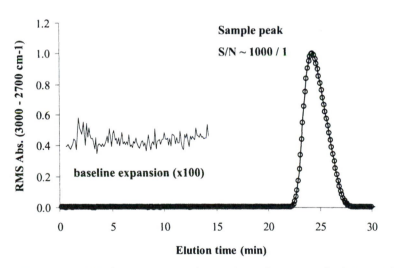

Figure 6. RMS chromatograms for a solvent fractionated ethylene 1-hexene copolymer. The insert illustrates the baseline noise during the first 15 minutes of the run.

The detection limit of a sample in chromatographic systems is dependent on eluent peak width, which in turn, will vary with flow rate, chromatographic resolution, and sample polydispersity. For example, the signal to noise dependence on a sample's polydispersity index (PDI) is illustrated in Table I. In the best case a S/N of 1,000/1 is obtained for a narrow MW PE fraction. Conversely, the S/N is 150/1 for a very broad resin (PDI ~ 70) made from a modified chromium oxide catalyst *(24)*.

TABLE I

Samples	Resin Type[a]	Nominal Signal/Noise[b]	Nominal PDI
A	Solvent fractionated PE	1,000/1	1.2
B	NBS 1475	450/1	3.7
C	Chromium oxide catalyzed resin	300/1	16
D	Modified, chromium oxide catalyzed resin	150/1	70

a) Using: 1.5 mg/mL sample in TCB, a 500 μL injection volume and flow rate = 1mL/min. at 140 °C on three PLgel Mixed A columns.

b) Using the sample peak max.

Because of the dynamic nature of this analysis, signal averaging is not an applicable technique to improve the signal to noise. Of course signal to noise can always be improved by narrowing the spectral window. For example, if the spectral window is changed from $3000 - 2700$ cm^{-1} to $2836 - 2920$ cm^{-1}, a significant increase in the signal to noise results (Figure 7). However, since the absorbance of this peak will change with branching content (i.e., less CH_2 moieties) the comparable concentration/detector responses for the various sample types may be compromised. Furthermore, obtaining separate spectral data and concentration data for each time slice from which the SCB and MW information are quantified, unnecessarily complicates the data processing portion of the analysis.

Typical molecular weight values obtained using SEC-FTIR under SCB analysis conditions are illustrated in Table II for the ethylene 1-olefin samples previously described in Table I. For SCB studies conducted using on-line SEC-FTIR *(11,15-18)*, large injection volumes (400–1000 μL) at sample concentrations between 1.0 and 3.0 mg/mL have been used in order to obtain sufficient amounts of sample for FTIR analysis.

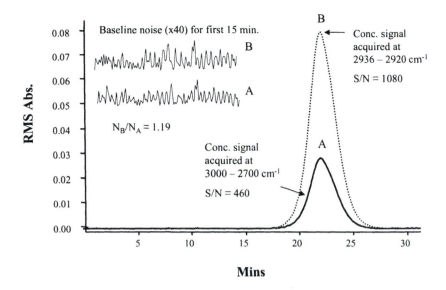

Figure 7. RMS chromatograms acquired for NBS 1475 (same run) using two different spectral windows. Noise for curves A and B were 6.4 x10⁻⁵ and 7.6 x10⁻⁵ RMS absorbance units, respectively (insert).

TABLE II

Sample [a]	SEC-FTIR			SEC-DRI [b]		
	Mn (kg/mol)	Mw (kg/mol)	PDI	Mn (kg/mol)	Mw (kg/mol)	PDI
A[c]	98.0	120	1.22	108	121	1.13
B	15.4	56.7	3.68	16.7	55.5	3.32
C	14.5	227	15.6	14.6	231	15.9
D	6.91	458	66.3	7.48	492	65.8

a) General chromatographic conditions as in Table I unless noted.

b) Using 220 uL injection volumes on three PLgel Mixed B columns, different SEC unit.

c) 1.1 mg/mL sample concentration *(11)*.

d) Molecular weight calculations were made using a broad molecular weight polyethylene (PE) standard *(11)*.

Because such high injection volumes and sample concentrations are typically used, one concern that arises is that column overloading may occur (i.e., the column is no longer separating the sample based on solely molecular weight), especially in samples with narrow MWDs or very high MWs. When this occurs, a general shift in molecular weight values to lower values can occur (i.e., lower M_n).

Detecting and Quantifying SCB Heterogeneity

The aim in any on-line SEC-FTIR method for branch analysis in polymers is to obtain SCB information over as much of the molecular weight distribution as possible using a predictive model for SCB/1000 Total Carbons (TC) and the appropriate software to generate values per chromatographic slice Some of the more pragmatic aspects of this method are addressed in this section.

Quantifying the branching levels in polyolefins using on-line SEC-FTIR requires a calibration curve that correlates the same portion of the sample's C-H stretching absorptions to its known methyl content (Me/1000 TC) as measured by NMR. In some studies (15,16,18), peak deconvolution and/or ratio methods are used which focus on the methyl and methylene peaks at 2958 cm^{-1} and 2928 cm^{-1}, respectively. In other methods, more of the spectral region is used coupled with multivariate analysis techniques (11,17). Most, if not all, of these methods are adequate if the methyl content in the sample is high enough as in the case of n-alkanes samples, high-pressure low-density polyethylene, poly (1-olefins), and some ethylene 1-olefins. However, in high-density samples (ρ > 0.94 g/cc) quantifying the methyl absorption is difficult and often limits the applicability of a particular method. In these cases, multivariable statistical techniques for data analysis have proved useful (11).

When assessing SCB heterogeneity across a sample's molecular weight distribution using on-line SEC-FTIR a number of considerations must be kept in mind. For example, changes in optical energy during the analysis may lead to error in the measured SCB values. This happens when the MCT detector has not completely cooled or when the level of liquid nitrogen coolant changes. The subsequent spectra are acquired at different optical energy outputs than was the starting background spectrum. When this happens, absorbance values across the spectra will change and the measured SCB level can vary by two SCB/1000 TC or more. Another source of error can occur in spectra collected at the tails of the chromatograph where the sample concentration is low. These spectra are difficult to use for SCB measurements due to poor signal to noise (S/N) ratios. Problems are usually encountered when S/N values of 20/1 are reached, which corresponds to ~0.065 mg/mL at the detector (11). This latter concentration can be considered the quantitative limit for this method.

Another concern is the proper selection of the number of methyl chain ends for a particular molecular weight slice when calculating a polyolefin's SCB content from methyl content. If an inaccurate level is used, then the estimated SCB content and therefore the perceived compositional heterogeneity in the low end of the molecular weight distribution (M_n < 15,000 g/mol) can be dramatically affected *(11)*. The proper choice for the number of methyl end groups is always problematic and a prior knowledge of the catalyst type used to produce the polymer is very helpful. For example, chain termination in chromium oxide catalyzed resins results in the formation of a one-methyl end group and one vinyl moiety *(25)*. Conversely, two methyl chain ends are formed in the chain termination step of Ziegler-Natta catalyzed resins *(26)*. Of course, the assigned values for the number of methyl chain ends for the two catalyst types are only generalizations. The number of methyl chain ends observed for polymers made from these and other catalysts fall between these two extremes (i.e., 1 and 2) and is dependent on the reaction conditions. For example, there are typically some FTIR detectable amounts of vinyl moieties in most ZN resins which means the value for V_{ce} > 0 (or Me_{ce} < 2). Conversely, if hydrogen is used in some modified chromium oxide catalyzed systems V_{ce} < 1 as is the case for resins made from Cr/aluminophosphate (V_{ce} ~ 0.8 and Me_{ce} ~ 1.2) *(24)*.

Further problems arise when a polymer is made from mixed catalyst types, as is the case in dual catalyst reactions or physical extruder blends. In these systems, the chain end composition of the final product at any one molecular weight will depend upon the relative weight fraction of each polymer type in the analyzed slice. Some prior knowledge of the resulting chain end composition and the expected resolution of the molecular weight distributions for the polymers produced from the catalysts are essential to adequately characterize these systems. This usually involves estimating the number of vinyl chain ends in the bulk sample using FTIR prior to the SEC-FTIR characterization.

The presence of unresolved impurities can also influence the SCB levels measured at the low end of the molecular weight distribution. A large increase in the measured SCB level can results when an antioxidant (AO) peak, caused from mismatched AO levels in the TCB sample diluent and mobile phase, overlaps with portions of chromatographed sample. This particular situation is illustrated in Figure 8 where the AO and n-alkane peaks are not well resolved. Since the spectrum of an AO such as butylated hydroxy toluene (BHT) has very strong methyl absorptions, dramatic changes in the measured SCB can occur even in those cases where only a slight overlap exists. As shown in Figure 8, the measured methyl content goes from the expected 55.6 Me/1000 TC to values >300 Me/1000 TC (of course this later value is only a gross estimate). Lastly, other sources of error associated with sampling heterogeneity, insolubility caused by high MW components or improper preparative conditions can influence how much of the HMW component is seen.

222

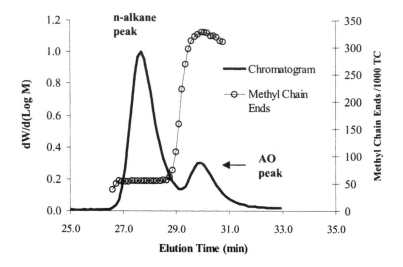

Figure 8. RMS chromatograms for an n-alkane (hexatriacontane) and the AO peak that arises from mismatched AO levels in the TCB sample diluent and mobile phase.

Compositional Heterogeneity Detected by SEC-FTIR

In the last section of this chapter, the extent to which SEC-FTIR detects compositional heterogeneity is assessed by examining the characterization of various polyolefin resins. Primary structural variations in polyolefins, and heterogeneity in the molecular weight distributions, will vary with the different catalyst types and conditions used to produce these polymers. For example, polyolefins made with multi-site catalysts, such as chromium oxide or Ziegler-Natta type catalysts, can produce polymers with PDIs > 4 with greater amounts of SCB in the low end of the molecular weight distribution. Typical SEC-FTIR characterizations of these resins are shown in Figure 9 However, other ZN catalyzed resins can exhibit more homogeneous SCB profiles if low amounts of comonomer are used in the reaction (Figure 10). In addition, other catalyst systems can also produce flat SCB profiles across the molecular weight distribution as determined by SEC-FTIR. Resins exemplifying this type of SCB profile include those made using modified chromium oxide *(24)* and certain types of metallocene catalysts (Figure 11).

The ability of SEC-FTIR analysis to discern a sample's compositional heterogeneity is dependent upon the extent to which the MWD of its components overlap. Therefore, in some cases, SEC-FTIR cannot fully elucidate a sample's compositional heterogeneity, since this method measures the *average* branch content in each slice of the molecular weight distribution. When the components of a polymer have similar molecular weights, SEC-FTIR analysis cannot detect compositional heterogeneity and other techniques must be used (11,13).

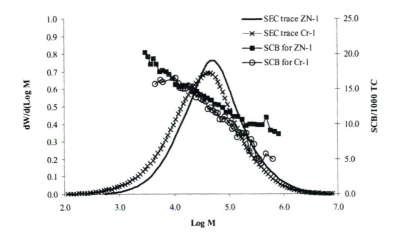

Figure 9. Comparison of comonomer incorporation in Ziegler-Natta (ZN-1) and chromium oxide catalyzed (Cr-1) ethylene 1-hexene resins.

One such technique typically used to assess the heterogeneity in polyolefin resins is analytical temperature rising elution fractionation (ATREF) *(3,27-29)*. Unlike the SEC-FTIR method, which separates polymer molecules by their size and calculates the average comonomer content within each slice of narrow molecular weight distribution (MWD), ATREF separates polymer molecules mainly by their crystallizability, which in turn is predominantly influenced by the comonomer content and its distribution within the molecules. This method yields a normalized distribution profile, of weight fraction vs. dissolution temperature, which reveals the comonomer composition of the resin. The spread of the temperature range and the variation of the weight fractions at different dissolution temperatures reflect the heterogeneity of the comonomer composition of the resin. Figure 12 illustrates an ATREF characterization of sample MTE-1.

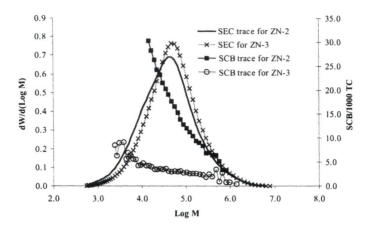

Figure 10. Comparison of comonomer incorporation in Ziegler-Natta catalyzed ethylene 1-hexene resins using high (ZN-2) and low (ZN-3) comonomer levels.

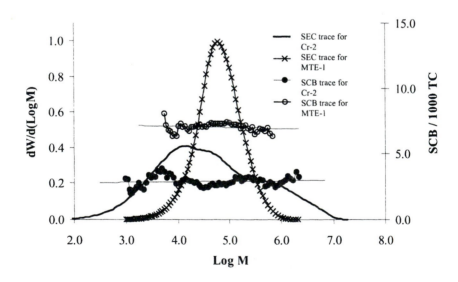

Figure 11. Comparison of comonomer incorporation in modified chromium oxide (Cr-2) and metallocene (MTE-1) catalyzed ethylene 1-hexene resins

The inability of SEC-FTIR analysis to discern a sample's compositional heterogeneity as the overlap between components increases is evident from a re-examination of Figures 11 and 12. For example the sample MTE-1 shown in Figure 11, was made by making a 50/50 blend of two different metallocene catalyzed resins of the same molecular weight and molecular weight distributions, but different branching content. The SCB profile of MTE-1 measured by on-line SEC-FTIR gives no indication that this polymer is a blend. By contrast, the corresponding ATREF profile of MTE-1 (Figure 12) clearly shows that it is composed of two distinct populations of polymeric molecules of very different comonomer content. However, it should be noted that ATREF would have difficulty discriminating compositional heterogeneity in resin blends that are composed of two or more components with similar comonomer compositions, but different molecular weights. For these latter types of resin blends, SEC-FTIR is the more appropriate method to detect compositional heterogeneity.

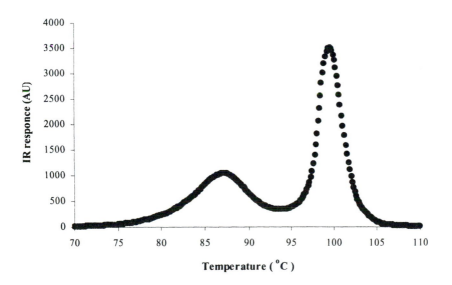

Figure 12. ATREF analysis of MTE-1(30).

Lastly, current on-line SEC-FTIR methods that use the stretching absorptions cannot distinguish certain types of topological heterogeneities. For example, ethylene 1-butene and ethylene 1-hexene copolymers with identical molecular weights, molecular weight distributions, and densities were characterized by SEC-FTIR . The SEC-FTIR analysis (Figure 13) suggests that both metallocene-catalyzed resins have homogeneous distributions of comonomers. In other words, these two resins look indistinguishable. However, the ATREF analysis of these resins shows them to be different and indicates that the comonomer distribution in the ethylene 1-hexene sample is more heterogeneous (Figure 14).

Chapter Summary

Commercial hardware and software developments over the last five years have made FTIR a viable on-line detector for high temperature SEC. In particular, on-line SEC-FTIR provides a rapid and accurate method to detect the compositional heterogeneity in many types polyolefins that result from catalyst and process changes. Data that took months to acquire, using preparative fractionation and NMR analysis, now can be obtained in less than one hour using on-line SEC-FTIR. However, the method has its limitations.

One obvious limitation for the on-line SEC-FTIR method is the fact that the presence of residual absorption from the mobile phase may obscure particular regions of the spectrum. Moreover, in the technique described in this chapter, methyl groups from both chain ends groups and short chain branches are included in the analyses. This latter fact makes both the estimation of chain ends, and proper column calibration, critical in accessing compositional heterogeneity. In addition, no distinction is made between different branch types in the stretching region of the spectrum. Furthermore, SCB information at the ends of the molecular weight distribution is not captured by this method.

The ability of on-line SEC-FTIR to discern a sample's compositional heterogeneity is dependent upon the extent to which the molecular weight distributions of its components overlap. In some cases, SEC-FTIR cannot fully elucidate a sample's compositional heterogeneity, and other techniques such as analytical temperature rising elution fractionation (ATREF) must be used. Clearly, the two methods look at the resin molecular architecture from entirely different perspectives and are complementary to one another. The choice of which method to use will be determined by the specific structure vs. property issue under study.

In light of the method's existing limitations, future work on improving on-line SEC-FTIR should focus on improving detector stability. Further improvements in commercial spectral-software, which allow for consistent batch analysis are also needed. Lastly, coupling this method with ATREF or light

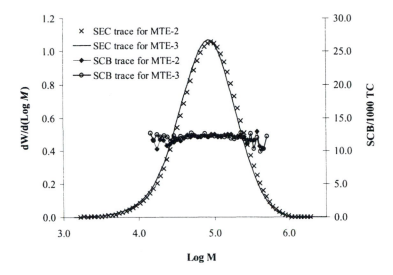

Figure 13. SEC-FTIR of the ethylene 1-butene and 1-hexene copolymers.

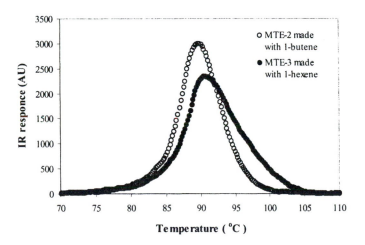

Figure 14. ATREF analysis of the ethylene 1-butene and 1-hexene copolymers shown in Figure 13 (30).

228

scattering techniques could enhance the effectiveness of on-line SEC-FTIR to detect compositional heterogeneity in polyolefins and other resins.

References

1. DesLauriers, P. J.; Battiste, D. R. *ANTEC-SPE* **1995**, 53(3), 3639.
2. Cady, L. D. *Plast. Eng.* January, 1987, p 25.
3. Hosoda, S. *J. Polym.* **1988**, 20, 383.
4. Zhang, M.; Lynch, D. T.; Wanke, S. E. *J. Appl. Polym. Sci.* **2000**, 75, 960.
5. Zhang, M.; Lynch, D. T.; Wanke, S. E. *Polymer* **2001**, 42, 3067.
6. Blitz, J. P., and McFadden, D. C. *J. Appl. Pol. Sci.* **1994**, 51,13.
7. *Annual book of ASTM Standards*; Standard D6645-01; American Society for Testing Materials: Philadelphia, PA, **2003**,Vol. 08.03.
8. Fox, J.J.; Martin, A.E. *Proc. Roy. Soc. (London)* **1937**,162, 419.
9. Wiberley, S. E.; Bruce, S. C.; Bauer, W. H. *Anal. Chem,* **1960**, 32, 217.
10. Bellamy, L. J.; *The Infra-red Spectra of Complex Molecules*, 3rd ed.; John Wiley & Sons: New York, NY, 1975; Ch. 2.
11. DesLauriers P. J.; Rohlfing, D. C.; Hsieh, E. T. *Polymer* **2002**, 43, 159.
12. Willis, J.N.; Wheeler, L. In *Chromatographic Characterization of Polymers: Hyphenated and Multidimensional Techniques,* Provder, T.;. Barth, H. G; Urban, M. W.; Marek , W.; Eds.; *Adv. Chem. Ser.* **1995**, 247.
13. Faldi, A.; Soares, J.B.P. *Polymer* **2001**, 42, 3057.
14. Somsen, G.W.; Gooijer, C.; Brinkman, U.A.Th. *J. Chromatogr. A* **1999**, 856, 213.
15. Housaki, T.; Satoh, K.; Nishikida, K.; Morimoto, M.; *Makromol Chem, Rapid Commun.* **1988**, 9, 525.
16. K. Nishikida, T. Housaki, M. Morimoto, and T. Kinoshita, *J Chromatography* **1990**, 517, 209.
17. Markovich RP, Hazlitt LG, Smith-Courtney L. In *Chromatography of Polymers: Characterization by SEC and FFF,* Provder, T.; Ed.; *ACS Symposium Series* **1993**, 521.
18. Dhenin, V.; Rose, L.J. *Polymer Preprints* **2000**, 41(1), 285.
19. Perkin Elmer Timebase software and software for measuring SCB codeveloped by Polymer Laboratories and Chevron Phillips Chemical Company and marketed by Polymer Laboratories.
20. A high temperature flow cell and codeveloped by Polymer Laboratories Chevron Phillips Chemical Company and marketed by Polymer Laboratories.
21. Nicolai, A.; Lederer, K. *Int. J. Polym. Anal. Charact.* **1997**, 3, 219.
22. Griffiths, P. R.; De Haseth, J. A. In *Fourier Transform Infrared Spectrometry;* Elving, P. J. and Winefordner, J. D. Eds.; *Chemical Analysis Series,* John Wiley and Sons. New York. NY, 1986, Vol.83, Ch. 6.
23. Vickrey, T.M. In *Liqid Chromatography Detectors;* Vickrey, T.M., Ed.; *Chromatographic Science Series,*Marcel Dekker: New York, NY, 1983, Vol 23, Ch. 1.
24. McDaniel, M. P.; Benham, E. A.; Wolfe, A. R.; DesLauriers, P. J. U.S. 6525148, 2003.

25. Witt, D.R. In *Reactivity, Mechanism and Structure in Polymer Chemistry* Jenkins, A.D.; Ledwith, A. Eds.;, John Wiley and Sons. New York. NY, 1974, Ch 13.
26. Lenz, R.W. *Organic Chemistry of Synthetic High Polymers.* John Wiley and Sons, New York, NY, 1967, Ch 15.
27. Pigeon, M.G.; Rudin, A. *J. Appl. Polym. Sci.* **1994**, 51, 303.
28. Yau, W.W.; Gillespie, D. *ANTEC-SPE* **2001**, 59 (1), 7 .
29. Tso, C.C.; DesLauriers, P.J.; Chevron Phillips Chemical Company unpublished results.
30. Data supplied by C. C. Tso, Chevron Phillips Chemical Company using conditions given in Reference 28.

Chapter 14

Characterization of Polymer Heterogeneity by 2D Liquid Chromatography

Harald Pasch

German Institute for Polymers (Deutsches Kunststoff-Institut),
Schlossgartenstrasse 6, 64289 Darmstadt, Germany

Complex polymers are distributed in more than one direction of molecular heterogeneity. In addition to the molar mass distribution, they are frequently distributed with respect to chemical composition, functionality, and molecular architecture. For the characterization of the different types of molecular heterogeneity it is necessary to use a wide range of analytical techniques. Preferably, these techniques should be selective towards a specific type of heterogeneity. The combination of two selective analytical techniques is assumed to yield two-dimensional information on the molecular heterogeneity.

The present review presents the principal ideas of combining different chromatographic techniques in two-dimensional analysis schemes. Most promising protocols for 2D-LC refer to the combination of interactive chromatographic methods in the first dimension and size exclusion chromatography in the second dimension. The detailed analysis of functional homopolymers, segmented copolymers, and complex polymer blends will be discussed.

Introduction

State-of-the-art polymeric materials possess property distributions in more than one parameter of molecular heterogeneity. Copolymers, for example, are distributed in molar mass and chemical composition, while telechelics and macromonomers are distributed frequently in molar mass and functionality. It is obvious that n independent properties require n-dimensional analytical methods for accurate (independent) characterization of the different structural parameters of a polymer (1).

Multidimensional chromatography separations can be done in planar systems or coupled-column systems. Examples of planar systems include two-dimensional thin-layer chromatography (TLC) (2, 3), and 2D electrophoresis, where gel electrophoresis is run in the first dimension followed by isoelectric focusing in the second dimension (4-6). Hybrids of these systems where chromatography and electrophoresis are used in each spatial dimension were reported nearly 40 years ago (7). Belenkii et al. reported on the analysis of block copolymers by TLC (8,9).

In 2D column chromatography systems an aliquot from a column or channel is transferred into the next separation method in a sequential and repetitive manner. Storage of the accumulating eluent is typically provided by sampling loops connected to an automated valve. Many variations on this theme exist which use various chromatographic and electrophoretic methods for one of the dimensions. In addition, the simpler „heart cutting" mode of operation takes the eluent from a first dimension peak or a few peaks and manually injects this into another column during the first dimension elution process. A partial compilation of these techniques is given in Refs (8, 10-13).

The use of different modes of liquid chromatography facilitates the separation of complex samples selectively with respect to different properties such as hydrodynamic volume, molar mass, chemical composition, architecture or functionality. Using these techniques in combination, multi-dimensional information on different aspects of molecular heterogeneity can be obtained. If, for example, two different chromatographic techniques are combined in a "cross-fractionation" mode, information on chemical composition distribution (CCD) and molar mass distribution (MMD) can be obtained. Literally, the term "chromatographic cross-fractionation" refers to any combination of chromatographic methods capable of evaluating the distribution in size and composition of copolymers. An excellent overview of different techniques and applications involving the combination of SEC and gradient HPLC was published by Glöckner in (14).

In the SEC mode, the separation occurs according to the molecular size of a macromolecule in solution, which is dependent on its chain length, chemical composition, solvent and temperature. Thus, molecules of the same chain length but different composition may have different hydrodynamic volumes. Since SEC separates according to hydrodynamic volume, SEC in different eluents can separate a copolymer in two diverging directions. This principle of "orthogonal chromatography" was suggested by Balke and Patel (15-17).

Since "orthogonality" requires that each separation technique is totally selective towards an investigated property, it seems to be more advantageous to use a sequence of methods, in which the first dimension separates according to chemical composition. In this way quantitative information on CCD can be obtained and the resulting fractions eluting from the first dimension are chemically homogeneous. These homogeneous fractions can then be analyzed independently in SEC mode in the second dimension to get the required MMD information. In such cases, SEC separation is strictly separating according to molar mass, and quantitative MMD information can be obtained.

Experimental Aspects of 2D-LC

Setting up a 2D chromatographic separation system is actually not as difficult as one might think at first. As long as well-known separation methods exist for each dimension the experimental aspects can be handled quite easily in most cases. Off-line systems just require a fraction-collection device and something or someone who reinjects the fractions into the next chromatographic dimension. In online 2D systems the transfer of fractions is preferentially done by automatic injection valves as was proposed by Kilz et al. (13, 18, 19). Figure 1 shows a general setup for an automated 2-dimensional chromatography system.

The system is composed of two chromatographs, including pumps, injector and column for the first dimension, and a pump, column and detectors for the second dimension. The focal point in 2D chromatography separations is the transfer of fractions eluting from the first dimension into the second dimension. This can be done in various ways. The most simplistic approach is collecting fractions from one separation and manually transfering them into the second separation system. Obviously, this approach is prone to many errors, labor intensive and quite time-consuming.

A more efficient way of fraction transfer can be achieved by using electrically (or pneumatically) actuated valves equipped with two injection

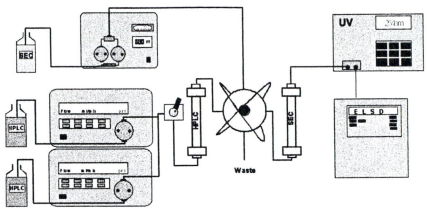

Figure 1. General experimental set-up for a 2D chromatographic system

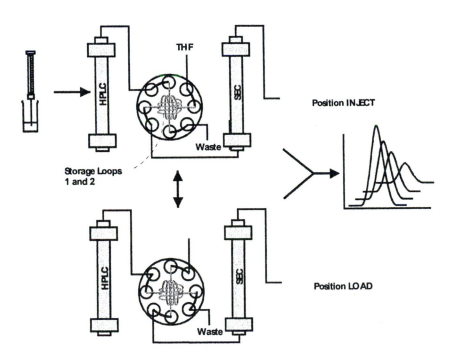

Figure 2. Fraction transfer between chromatographic dimensions using a dual loop 8-port valve

loops. Such a set-up allows one fraction to be injected and analyzed from one loop while the next fraction is collected at the same time in the second loop. The operation of such an automatic dual-loop system is schematically presented in Figure 2. The operation of the system is explained in detail in Ref. (20).

There are some other important aspects which have to be considered for optimum 2D experiment design, including the selection of separation techniques, the sequence of separation methods, and the detectability in the second dimension. The sequence of separation methods is an important aspect in order to obtain the best resolution and most accurate determination of property distributions. It is advisable to use the method with highest selectivity for the separation of one property as the first dimension. This ensures the highest purity of eluting fractions being transfered into the subsequent separation. In the case of gradient HPLC and SEC as separation methods, early publications (15-17, 21, 22) used SEC as the first separation, because it took much longer than subsequent HPLC analyses. This is not the best set-up, however, because the SEC fractions are only monodisperse in hydrodynamic volume, but not in molar mass, chemical composition, etc. On the other hand, HPLC separations can be fine-tuned using gradients to fractionate only according to a single property, which can then be characterized for molar mass without any bias.

In many cases, interaction chromatography as the first dimension separation method is the best and most adjustable choice. From an experimental point of view, high flexibility is required for the first chromatographic dimension. In general, this is also achieved more easily when running the interaction chromatography mode as the first dimension, because (a) more parameters (mobile phase composition, mobile phase modifiers, stationary phase, temperature, etc.) can be used to adjust the separation according to the chemical nature of the sample, (b) better fine-tuning in interaction chromatography allows for more homogeneous fractions, and (c) sample load on such columns can be much higher as compared to SEC columns.

Separation Techniques for the First and Second Dimensions

In 2D chromatography, different modes of liquid chromatography are combined with each other. Depending on the individual technique, separations can be carried out with regard to molecular size, chemical composition, or architecture.

The most often used set-up for 2D chromatography is the combination of interaction chromatography and size exclusion chromatography. SEC is the standard technique for determining molar mass distributions by separating according to hydrodynamic volume. For complex polymers it must be considered, however, that hydrodynamic volume is not only a function of chain length (molar mass) but also of chemical composition and architecture. It is, therefore, not selective towards only one parameter of molecular heterogeneity.

Interaction chromatography, on the other hand, can be performed in a large variety of different experimental set-ups. Normal phase or reversed phase systems using isocratic or gradient elution can be used. There is an abundance of stationary phases with different types of surface modifications of different polarities. This flexibility in experimental parameters is a very important criterion for using interaction chromatography as a first dimension method, since it can be fine-tuned to separate according to a given property more easily than most other chromatographic techniques.

Gradient HPLC has been useful for the characterization of copolymers (23-27). In such experiments careful choice of separation conditions is imperative. Otherwise, low resolution for the polymeric sample will obstruct the separation. On the other hand, the separation in HPLC, dominated by enthalpic interactions, perfectly complements the entropic nature of the SEC retention mechanism in the characterization of complex polymer formulations.

Another mode of interaction chromatography - liquid chromatography at critical conditions (LCCC) - relates to a chromatographic situation where the entropic and enthalpic interactions of the macromolecule and the column packing compensate each other. The Gibbs free energy of the macromolecule remains constant when it penetrates the pores of the stationary phase. The distribution coefficient K_d is unity, regardless of the size of the macromolecules, and all macromolecules of equal chemical structure elute from the chromatographic column as one peak. Molecules of other chemical compositions are off-critical and elute at conditions that are influenced by MMD. The term "chromatographic invisibility" is used to refer to this phenomenon. This means that the chromatographic behaviour is not directed by the size, but by the heterogeneities (chemical structure, architecture, end groups, etc.) in the macromolecular chains (28-31). This chromatographic effect can be employed to selectively determine imperfections in the polymer chain, without any contribution from the repeat units themselves. LCCC has been successfully used for the determination of the functionality type distribution of telechelics and macromonomers, and for the analysis of block copolymers, macrocyclic polymers, and polymer blends (20).

Although much of the early work on multidimensional chromatography was focused on using SEC in the first dimension, it is now widely accepted that an interactive type of separation should be used as the first step. On the other hand, SEC is preferentially used as the second method to retrieve molar mass information.

Coupling of Gradient HPLC and SEC

Much work on chromatographic cross-fractionation was carried out with respect to the combination of SEC and gradient HPLC. In most cases SEC was used as the first separation step, followed by HPLC. In a number of early papers the cross-fractionation of model mixtures was discussed. Investigations of this kind demonstrated the efficiency of gradient HPLC for separation by chemical composition. Mixtures of random copolymers of styrene and acrylonitrile were separated by Glöckner et al. (22). In the first dimension a SEC separation was carried out using tetrahydrofuran (THF) as the eluent and polystyrene gel as the packing. In total, about 10 fractions were collected and subjected to the second dimension, which was gradient HPLC on a CN bonded phase using iso-octane/THF as the mobile phase. Model mixtures of random copolymers of styrene and 2-methoxyethyl methacrylate were separated in a similar way, the mobile phase of the HPLC mode being iso-octane/methanol in this case (32). This procedure was also applied to real-world copolymers (22). Graft copolymers of methyl methacrylate onto EPDM rubber were analyzed by Augenstein and Stickler (33), whereas Mori reported on the fractionation of block copolymers of styrene and vinyl acetate (34). In all these experiments the same limitation with respect to the SEC part holds true: when SEC is used as the first dimension, true molar mass distributions are not obtained.

From the theoretical point of view, a more feasible way of analyzing copolymers is the prefractionation through HPLC in the first dimension and subsequent analysis of the fractions by SEC (35, 36). HPLC was found to be rather insensitive towards molar mass effects and yielded very uniform fractions with respect to chemical composition.

As a representative example the following application describes the analysis of the grafting reaction of methyl methacrylate onto EPDM (37). Graft copolymers with an elastomeric back bone and thermoplastic grafts serve as impact modifiers in rubber-modified thermoplastics. This type of graft copolymer A-g-B (e.g. EPDM-g-PMMA) is prepared by polymerizing a monomer B (e.g. MMA) radically in the presence of a polymer A (e.g. EPDM). Grafts B grow from macroradicals A.

As a result of the grafting reaction a complex product is obtained comprising the graft copolymer A-g-B, residual ungrafted polymer backbone A (e.g. PMMA) and homopolymer B. Accordingly, the reaction product is distributed both in molar mass (MMD) and chemical composition (CCD).

As a result of the grafting reaction a complex product is obtained comprising the graft copolymer A-g-B, residual ungrafted polymer backbone A (e.g. PMMA) and homopolymer B. Accordingly, the reaction product is distributed both in molar mass (MMD) and chemical composition (CCD).

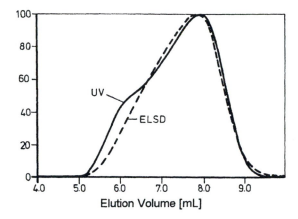

Figure 3. SEC chromatogram of the graft product sample 2; stationary phase: PSS SDV linear, eluent: THF, detection: (——) UV 254 nm, (- - -) ELSD (Reproduced from Ref. (37). Copyright 2001 Wiley-VCH, Germany).

Initially, size exclusion chromatography (SEC) is used to separate with respect to molecular size. The SEC chromatogram of a typical reaction product is shown in Fig. 3. It indicates that the molar mass differences of the components are not large enough for separate elution peaks. However, an indication on the distribution of methyl methacrylate across the elution volume axis can be obtained by comparing the traces of the evaporative light scattering detector (ELSD) and UV (254 nm) detectors. Since PMMA has a higher UV response at 254 nm than EPDM, it becomes apparent that the lower elution volume (higher molar mass) part of the chromatogram is rich in PMMA. This part presumably belongs to the graft copolymer, however, definite information cannot be obtained due to incomplete separation.

One very useful technique for the separation of copolymers according to chemical composition is gradient HPLC. The combined effects of adsorption and precipitation in high performance precipitation LC can be used to separate complex reaction products into their respective components. In the case of EPDM-*g*-PMMA a polar Nucleosil CN stationary phase is used for gradient HPLC. A stepwise gradient of THF/i-octane is used starting with 99 % by volume of i-octane and going to 100 % by volume THF. The resulting chromatograms of three samples are presented in Fig. 4. A perfect separation into three fractions is obtained for all samples. Samples 2, 6, and 9 are reaction products that were taken after 150, 200, and 420 min from the reaction mixture.

238

The assignment of the peaks is carried out by comparison with the chromatographic behaviour of EPDM and PMMA, assuming that the elution order is a function of component polarity. As the least polar component EPDM is eluted first, followed by the graft copolymer EPDM-*g*-PMMA, with the most polar PMMA homopolymer last. The small peak at a retention time of 5 min is due to the fact that EPDM is not fully retained on the column after injection. At full retention of EPDM this peak does not appear in the chromatograms.

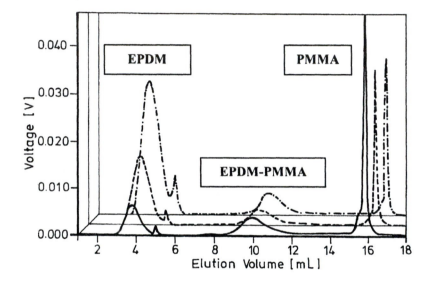

Figure 4. Gradient HPLC chromatograms of graft products samples (-.-.-.) 2, (----)6, and (——) 9; stationary phase: Nucleosil CN 300+500Å, eluent: THF/i-octane, detection: ELSD (Reproduced from Ref. (37). Copyright 2001 Wiley-VCH, Germany).

While the gradient HPLC experiments yield detailed information on the chemical composition of the graft products, information on molar masses must be obtained by SEC. For an optimization of the conditions of the grafting reaction it is of particular interest to determine the molar mass distribution of each of the product components. Using on-line coupled 2D chromatography dual information on chemical composition and molar mass can be obtained. In the first chromatographic dimension separation is conducted with respect to chemical composition using gradient HPLC, while in the second dimension SEC separation is carried out with respect to molar mass.

The results of the two-dimensional separation of one graft product is presented as a contour diagram in Plate 1. The ordinate represents the separation in the first dimension, while the abscissa indicates the results of SEC separation of the fractions. The contour plot clearly indicates that the molar masses of the three components are of similar magnitude. Therefore, an SEC-type separation alone (as in Fig. 3) could obviously not resolve the different components of the graft product. The relative concentrations of the components are obtained from the intensities of the contour plot peaks. For sample 2 EPDM has the highest concentration, while the concentrations of the reaction products EPDM-g-PMMA and PMMA are comparable. For a sample taken at a later stage of the grafting reaction the relative concentration of EPDM is much lower and PMMA constitutes the major product.

Coupling of Liquid Chromatography at the Critical Point of Adsorption and SEC

As already mentioned, liquid chromatography at the critical point of adsorption (LCCC) relates to a chromatographic situation where the entropic and enthalpic interactions of the macromolecules and the column packing compensate each other perfectly. Under such chromatographic conditions it is possible to determine the heterogeneities of the polymer chain selectively and without influence of the polymer chain length. Thus, LCCC represents a chromatographic separation technique yielding fractions which are homogeneous with respect to chemical composition but distributed in molar mass. These fractions can readily be analysed by SEC which, for chemically homogeneous fractions, provides true molar mass distributions without interference of CCD or functionality type distribution (FTD). Therefore, the combination of LCCC and SEC in a 2D set-up can truly be regarded as "orthogonal" chromatography provided that LCCC comprises the first dimension. Consequently, for functional homopolymers which are distributed in both functionality and molar mass, coupled LCCC vs. SEC can yield combined information on FTD and MMD.

Epoxy resins are one of the most important types of cross-linked polymers. High chemical corrosion resistance, good mechanical and thermal properties, and outstanding adhesion to various substrates are characteristics of these materials. Epoxy resins are usually prepared by the reaction of epichlorohydrin and bisphenol A (BPA). As a result of this reaction oligomers are formed which contain mainly glycidyl endgroups. Due to side reactions such as hydrolysis of the epoxy groups or incomplete dehydrohalogenation, other endgroups may also be formed. In subsequent reactions branching can also take place. Accordingly, epoxy resins may exhibit a functionality type distribution

240

and a topological distribution in addition to the usual molar mass distribution. 2D-LC is the perfect tool to analyze such complex samples (38, 39).

For the analysis of the functionality type distribution of the epoxy resins LCCC is used. The critical point of adsorption corresponds to a mobile phase composition of THF-hexane 74:26 % by volume, the stationary phase in this case is silica gel.

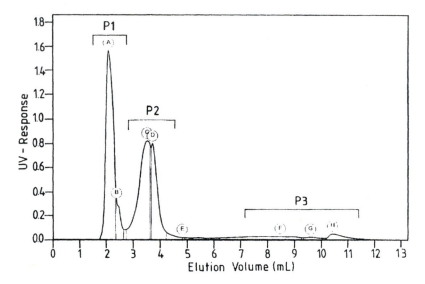

Figure 5. LCCC chromatogram of sample 8, stationary phase: Nucleosil 50-5, mobile phase: THF-hexane 74:26 % by volume, detection: UV 280 nm (Reproduced from Ref. (38). Copyright 2001 Wiley-VCH, Germany)

A typical LCCC chromatogram is given in Fig. 5. Sample 8 is a commercial epoxy resin with a number average molar mass of 1000 g/mol. Three well separated elution regions P1, P2, and P3 were obtained. The broadness of P2 and P3 indicates, however, that these fractions do not behave critically or they are still heterogeneous. For a detailed analysis of the fractions, sample 8 was fractionated multiple times to obtain the narrow fractions A-H, which were subjected to MALDI-TOF mass spectrometry (38). The MALDI-TOF assignment of the chromatographic fractions is given in Table I.

Table I. Assignment of epoxy resin structures to the fractions of sample 8 *

Fraction	General Structure
A,B	
C	
D	
F,G	
H	

* for fraction E no useful spectrum could be obtained

As can be seen, the first fractions contain the oligomers that are rich in epoxy groups. It is interesting that in addition to linear oligomers mono- and dibranched oligomers were identified. These oligomers obviously result from the reaction of a terminal epoxy group with the secondary hydroxyl group of another oligomer. The fractions C-G correspond to oligomers that contain epoxy and diol terminal groups. In contrast, fraction H belongs to oligomers without epoxy groups. Such oligomers cannot react in a cross-linking reaction and are, therefore, unwanted by-products.

Using the same chromatographic conditions, LCCC and SEC were combined in an on-line coupled two-dimensional set-up. Only the flow rates of both methods had to be adjusted for the 2D experiment. The two-dimensional experiment yields separation with respect to functionality and molar mass, and both the FTD and the MMD can be determined quantitatively. The contour plot of sample 8 in Plate 2 reveals a number of features not seen in the off-line LCCC and SEC experiments. Three regions, 1-3, are obtained in the contour plot, which correspond to P1, P2 and P3 in Fig. 6 However, these regions are not uniform but exhibit different substructures which are coded as (a), (b), and (c).

The substructures can be assigned to linear (a), monobranched (b) and dibranched oligomer series (c). Such structures have also been detected by MALDI-TOF-MS. As can be seen, the resolution of 2D chromatography far exceeds the resolution capabilities of LCCC and SEC. Not only are different functionality fractions observed but also different degrees of branching. In addition, for each oligomer series the quantitative oligomer distribution can be obtained. As was pointed out previously, the color code in the contour plot corresponds to the concentration of each species. Accordingly, for each structural feature the concentration and molar mass distribution can be obtained. The wealth of information obtained by a 2D-LC experiment is clearly demonstrated in Table II.

Further Applications and Outlook

As has been shown, two-dimensional liquid chromatography is one of the most powerful methods for characterizing complex polymers in different coordinates of molecular heterogeneity. Using a chromatographic separation which is selective towards functionality or chemical composition in the first dimension and SEC in the second dimension, truly "orthogonal" separation schemes can be established. Thus, the combination of gradient HPLC vs. SEC yields quantitative information on CCD and MMD, while coupling LCCC and SEC is useful for the analysis of functional homopolymers and block copolymers in the coordinates FTD-MMD and CCD-MMD, respectively. Even more complex systems, such as graft copolymers and polymer blends, in which each component may itself be chemically heterogeneous, can be analysed. Further applications of 2D chromatography to complex polymers are summarized Refs. (40-45).

Although 2D liquid chromatography is experimentally more demanding than other chromatographic techniques, the complete characterization yields much more qualitative and quantitative information about the sample, and results are presented in an impressively simple way. The contour plot of a 2D separation maps all obtainable information and allows a fast and reliable comparison between two samples. Regarding future development, the automated comparison of the results of different samples should be considered an important step toward improving process control and quality management. It can be expected that, in addition to LAC and LCCC, other separation modes will be combined with SEC. In an interesting application anion-exchange chromatography was coupled to SEC in order to analyse poly- and oligosaccharides (45).

Table II. Amounts and average molar masses of the functionality fractions of sample 8

Region	Structure	Amount [%]	M_n [g/mol]	M_w [g/mol]
1	*(structure)*	65	860	1720
2	*(structure)*	4	1690	2620
	(structure)	26	1130	1840
	Total	30	1180	1940
3	*(structure)*	1	2990	3680
	(structure)	1	1360	1590
	(structure)	3	1100	1990
	Total	5	1215	2250
Total		100	950	1810

References

1. Grushka, E., *Anal. Chem.* **1970**, *42*, 1142
2. Consden, R., Gordon, A.H., Martin, A.J.P. *Biochem. J.* **1944**, *38*, 224
3. Grinberg, N., Kalász, H., Han, S.M., Armstrong, D.W. In: *Modern Thin-Layer Chromatography* (Grinberg, N., Ed.) Marcel-Dekker Publishing, New York, 1990, Chapter 7
4. O'Farrell, P.H., *Biol. Chem.* **1975**, *250*, 4007

244

5. Celis, J.E., Bravo, R., *Two-Dimensional Gel Electrophoresis of Proteins*, Academic Press, New York, 1984
6. Anderson, N.L., Taylor, J., Scandora, A.E., Coulter, B.C., Anderson, N.G. *Clin. Chem.* **1981**, *27*, 1807
7. Efron, M.L., *Biochem. J.* **1959**, *72*, 691
8. Gankina, E., Belenkii, B., Malakhova, I., Melenevskaya, E., Zgonnik, V. *J. Planar Chromatogr.* **1991**, *4*, 199
9. Litvinova, L.S., Belenkii, B.G., Gankina, E.S. *J. Planar Chromatogr.* **1991**, *4*, 304
10. Murphy, R.E., Schure, M.R., Foley, J.P. *Anal. Chem.* **1998**, *70*, 1585
11. Balke, S.T. *Quantitative Column Liquid Chromatography*, Journal of Chromatography Library, Elsevier Publishing, New York, 1984, Vol. 29
12. Cortes, H.J. *Multidimensional Chromatography: Techniques and Applications,* Marcel Dekker, New York, 1990
13. Kilz, P., Krüger, R.P., Much, H., Schulz, G. In: *Chromatographic Characterization of Polymers: Hyphenated and Multidimensional Techniques*; Provder, T., Urban, M.W., Barth, H.G. (Eds.), Adv. Chem. Ser. 247; American Chemical Society, Washington, D.C., 1995
14. Glöckner, G. *Gradient HPLC and Chromatographic Cross-Fractionation,* Springer, Berlin-Heidelberg-New York, 1991
15. Balke, S.T., Patel, R.D. *J. Polym. Sci. B Polym. Lett.* **1980**, *18*, 453
16. Balke, S.T. *Sep Purif Methods* **1982**, *1*, 1
17. Balke, S.T., Patel, R.D. In: *Polymer Characterization*; Craver, C.D. (Ed), Adv. Chem. Ser. 203, American Chemical Society, Washington, D.C., 1983
18. Kilz, P. *Laborpraxis* **1992**, *6*, 628
19. Kilz, P., Krüger, R.P., Much, H., Schulz, G. *Polym. Mater. Sci. Eng.* **1993**, *69*, 114
20. Pasch, H., Trathnigg, B. *HPLC of Polymers*, Springer, Berlin-Heidelberg-New York, 1998
21. Ogawa, T., Sakai, M. *J. Polym. Sci. Polym. Phys. Ed.* **1982**, *19*, 1377
22. Glöckner, G., van den Berg, J.H.M., Meijerink, N.L., Scholte, T.G. In: *Integration of Fundamental Polymer Science and Technology*; Kleintjens, I., Lemstra, P. (Eds.), Elsevier Applied Science, Barking, 1986
23. Danielewicz, M., Kubín, M. *J. Appl. Polym. Sci.* **1981**, *26*,951
24. Glöckner, G., Koschwitz, H., Meissner, C. *Acta Polym.* **1982**, *33*, 614
25. Sato, H., Takeuchi, H., Tanaka, Y. *Macrommolecules* **1986**, *19*, 2613
26. Mourey, T.H. *J. Chromatogr.* **1986**, *357*, 101
27. Mori, S. *J. Appl. Polym. Sci.* **1989**, *38*, 95
28. Belenkii, B.G., Gankina, E.S., Tennikov, M.B., Vilenchik, L.Z. *Dokl. Akad. Nauk USSR* **1976**, *231*, 1147 (In Russian)
29. Tennikov, M.B., Nefedov, P.P., Lazareva, M.A., Frenkel, S.J. *Vysokomol. Soedin.* **1977**, *A19*, 657 (In Russian)

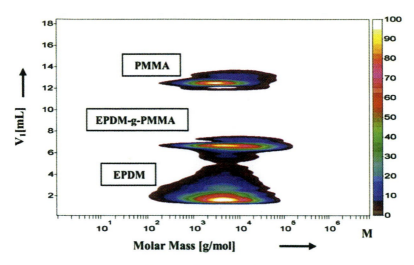

Plate 14.1. Contour plot of the two-dimensional separation of the graft product sample 2; 1st dimension: gradient HPLC, 2nd dimension: SEC, detection; ELSD (Reproduced from Reference 37. Copyright 2001 Wiley-VCH, Germany.)

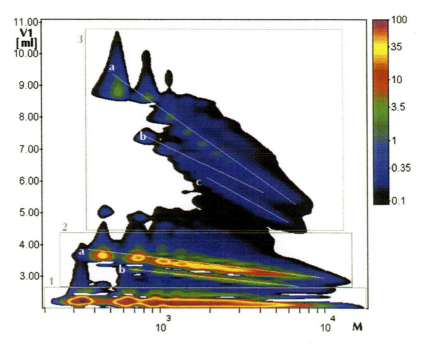

Plate 14.2. Contour plot of the 2D separation of sample 8; 1st dimension: LCCC, 2nd dimension: SEC, detection: UV 280 nm, regions 1, 2, 3 and series a, b, c see text (Reproduced from Reference 39. Copyright 2001 Wiley-VCH, Germany.)

30. Skvortsov, A.M., Belenkii, B.G., Gankina, E.S., Tennikov, M.B. *Vysokomol. Soedin.* **1978**, *A20*, 678 (In Russian)
31. Entelis, S.G., Evreinov, V.V., Gorshkov, A.V. *Adv. Polym. Sci.* **1986**, *76*, 129
32. Glöckner, G., Stickler, M., Wunderlich, W. *J. Appl. Polym. Sci.* **1989**, *37*, 3147
33. Augenstein, M., Stickler, M. *Makromol. Chem.* **1990**, *191*, 415
34. Mori, S. *J. Chromatogr.* **1990**, *503*, 411
35. Mori, S. *Anal. Chem.* **1988**, *60*, 1125
36. Mori, S. *Anal. Chem.* **1981**, *53*, 1813
37. Siewing, A., Schierholz, J., Braun, D., Hellmann, G., Pasch, H. *Macromol. Chem. Phys.* **2001**, *202*, 2890
38. Adrian, J., Braun, D., Rode, K., Pasch, H. *Angew. Makromol. Chem.* **1999**, *267*, 73
39. Adrian, J., Braun, D., Rode, K., Pasch, H. *Angew. Makromol. Chem.* **1999**, *267*, 82
40. Adrian, J., Esser, E., Hellmann, G., Pasch, H. *Polymer* **2000**, *41*, 2439
41. Pasch, H., Mequanint, K., Adrian, J. *e-Polymers* **2002**, No. 005
42. Biela, T., Duda, A., Penczek, S., Rode, K., Pasch, H. *J. Polym. Sci. A, Polym. Chem.* **2002**, *40*, 2884
43. Graef, S., van Zyl, A.J.P., Sanderson, R.D., Klumperman, B., Pasch, H. *J. Appl. Polym. Sci.* **2003**, *88*, 2530
44. van der Horst, A., Schoenmakers, P.J. *J. Chromatogr. A* **2003**, *1000*, 693
45. Suortti, T. *J. Chromatogr. A* **1997**, *763*, 331

Chapter 15

Composition Characterization of Polymers by 2D and Multiple-Detection Polymer Chromatography

Iwao Teraoka

Department of Chemical Engineering and Chemistry,
Polytechnic University, Brooklyn, NY 11201

This chapter explains novel two-dimensional chromatographic methods for characterizing polymers. The first dimension is either high osmotic pressure chromatography (HOPC) or phase fluctuation chromatography (PFC), where a concentrated solution of a given polymer is injected into a column until the latter is filled with solution. The second dimension is provided by various methods, including size-exclusion chromatography (SEC). Rather than maintaining a one-to-one relationship between the retention volume and the property of the polymer that elutes at that volume, HOPC and PFC pay attention to securing, at high resolution, as much of a polymer as possible in each fraction. Because each fraction has a large amount of polymer it can be characterized off-line by chromatographic, spectroscopic, and thermal analyses in the second dimension. The principles of separation at high concentrations are explained first, followed by examples of characterization of random copolymers, telechelic polymers, and block copolymers.

Introduction

Characterization of polymer in two or more dimensions often requires that the separation in the first dimension produces fractions containing a large amount of polymer — a few mg or more in each fraction. This requirement is not easily met by size-exclusion chromatography (SEC) in the first dimension, unless it uses a bank of large columns and, therefore, a large amount of solvent. We realized a decade ago that fully loading an analytical-size column with a concentrated solution of a given polymer permits the desired separation in the first dimension (1,2). The separation using a column packed with porous particles that have neutral walls is called *high osmotic pressure chromatography* (HOPC) (1-9). The result is separation primarily by molecular weight (MW). The separation using a column with a wall that gives different levels of attraction to the polymer components is called *phase fluctuation chromatography* (PFC) (10-16). In the latter, separation by chemical composition is intended. Since their inception in mid 1990s, HOPC and PFC have been applied to various polymers. The main purpose was preparative separation, but their capacity was also used for characterization. Table I lists examples of the two-dimensional separation and/or multiple-detection characterization using HOPC or PFC in the first dimension. The list is not exhaustive; other types of polymers can be analyzed and other methods can be used in the second dimension. Because the separation principle is different from the principles that govern SEC and HPLC, we first describe the principles of separation and then give a brief account of the examples.

Table I. Examples of Two-dimensional Separation

Polymer	First Dimension	Second Dimension[a]	Information Obtained
random copolymer	PFC	IR (11,14), DSC (10)	chemical composition distribution
block copolymer	PFC	NMR (12,13,15, 16), SEC (16), HPLC (15)	distributions in number of blocks and block length
telechelic polymer	HOPC	SEC (8,9), ^{1}H-NMR (8)	terminal chemistry distribution
homologous polymer	HOPC	SEC (1-9), MS (4)	MW distribution

[a]The numbers in parentheses denote references.

Separation Principles

Partitioning

All chromatographic separations are based on partitioning of a solution containing analyte molecules between a stationary phase and a mobile phase. In polymer separation the stationary phase is often identical to the space within the pores of the packing materials, and the mobile phase is the interstitial volume between the porous particles. Below, partitioning is explained separately for low concentrations and high concentrations.

Partitioning at Low Concentrations

The dashed line in Figure 1a is a sketch of the partition coefficient K of a polymer in dilute solutions with a pore of neutral walls as a function of MW of the polymer. By 'neutral' it is meant that the change in enthalpy, ΔH, upon transferring a polymer chain from the mobile phase to the stationary phase is zero. The monotonically decreasing K underlies SEC. As the pore surface turns attractive (relative to the solvent) the plot is lifted, especially at the high end of MW. When $-\Delta H$ per repeating unit of the polymer is sufficiently large, the plot of K is inverted and now becomes an increasing function of MW (solid line), because a longer chain has a greater $-\Delta H$ (17). The positive MW dependence of K can be used to separate oligomers by MW.

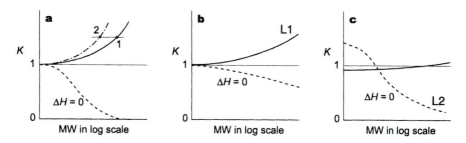

Figure 1. Partition coefficient K of a polymer, plotted as a function of MW. (a) Dilute Solutions. (b) Semidilute Solutions of a monodisperse polymer. (c) Semidilute Solutions of a polydisperse polymer.

Another line such as the one indicated by a dash-dotted line can be drawn in Figure 1a for a polymer with a different chemical composition and therefore a different ΔH per repeat unit. Along the solid line, point 1 indicates a specific MW for a given chemical composition. Along the dash-dotted line, point 2 refers to a polymer with another specific MW for another chemical composition. These two polymers have the same K, and therefore the column cannot distinguish them, although their chemical compositions are different. This is why separation of a polymer by chemical composition, independent of MW, has been a difficult problem in chromatographic characterization of polymers by isocratic elution (*18*).

Somewhere between the dashed line and the solid line we can expect an MW-independent K when size-exclusion is compensated for by the interaction (critical condition). The latter condition means, however, no separation, as $K = 1$ for all MW. As soon as ΔH turns negative to increase the polymer retention or positive to exclude the polymer, it is greeted by a strong MW dependence.

The critical condition depends on the mobile phase. A polymer of a different chemical composition is at the critical condition in a different mobile phase. It is then possible to utilize the MW-independent partitioning at the critical condition to separate heterogeneous polymers with respect to the chemical composition using gradient elution. A gradual change in the mobile phase composition, or temperature, with time will result in the release of one polymer of a specific composition after another. This gradient elution mode has been successfully used to characterize the chemical composition of polymers (*19-21*).

At low concentrations the partitioning of each polymer chain is independent from that of other chains. Therefore, the plots of Figure 1a apply to a series of dilute solutions, each containing a monodisperse polymer of a different MW, and to a solution of a polydisperse polymer. For the latter, K is defined for each MW component of the polymer.

Partitioning at High Concentrations

The partitioning is dictated by the thermodynamics of polymer solutions. Therefore, the partitioning at high concentrations of solutions containing monodisperse polymers is different than that of a solution of a polydisperse polymer. Let us first consider solutions of monodisperse polymers. It is known that in a solution of polymer in a good solvent the osmotic pressure deviates upward from van't Hoff's law with increasing concentration (*22*). Above the overlap concentration (where the solution viscosity is about twice as high as that of the solvent) and in the semidilute range, the deviation is quite large. The osmotic pressure of the semidilute solution can be a hundred times as high as that of an ideal solution of the same concentration. The deviation is particularly large for a high-MW polymer that has a low overlap concentration.

In the partitioning of such solutions with a pore of neutral walls, the upward deviation of the osmotic pressure occurs primarily in the surrounding solution where the concentration is higher compared with the solution in the pore ($K < 1$). Therefore, the polymer chains are driven into pores at a higher proportion than they are in the low concentration limit. The resultant increase in K causes the concentration in the pore to increase more rapidly than it does in the surrounding solution, but the concentration is always lower in the pore. The increase in K with an increasing concentration continues until K becomes close to one. The approach of K to unity requires a higher concentration for a higher-MW polymer, because the size-exclusion effect is stronger. As a result, the partition coefficient for a series of monodisperse polymers in solutions of the same concentration changes with their MW as depicted by a dashed line in Figure 1b. The MW dependence is subdued, compared with the dashed line in Figure 1a. For a theoretical formulation of the above, based on polymer solution thermodynamics, see references (23-26).

The smaller MW dependence in semidilute solutions compared with dilute solutions is also the case for attractive pores ($\Delta H < 0$). Now the concentration increases first within the pore. The higher osmotic pressure of the solution in the pore compared with the surrounding solution drives the polymer chains out of the pore. Thus, K decreases with increasing concentration. The MW dependence is shown as a solid line (L1) in Figure 1b for a series of solutions containing monodisperse polymers.

The plots of K for a polydisperse polymer in semidilute solution are different from those in Figure 1b. Firstly, we consider the partitioning with a neutral pore. When the surrounding solution has low- and high-MW components the high osmotic pressure pushes the low-MW components into the pore more than it does the high-MW components. Consequently, K for the low-MW component becomes higher than that in the solution of that component alone at the same concentration. The concentration may be even higher in the pore than it is in the surrounding solution ($K > 1$). The opposite is true for the high-MW component. In effect, the polymer is segregated by MW between the stationary phase and the mobile phase. The segregation cannot occur at low concentrations. The dashed line (L2) in Figure 1c illustrates the partition coefficient. In HOPC segregation of the polymer by MW is repeated at every theoretical plate in the column, thereby enriching the early eluent with high-MW components. For a theoretical formulation of this process, see references (27,28).

The partitioning depends on the MW distribution of the polymer. A polymer sample with a unimodal MW distribution and another sample with a bimodal distribution will exhibit different plots even if their weight-average and number-average MW are identical. The partitioning also depends on the concentration.

At the overlap concentration the plot of K should be close to the dashed line in Figure 1a. With increasing concentration the contrast of K between low- and high-MW components becomes stronger. With a further increase the contrast diminishes, as K approaches one for all MWs. Thus, there should be an optimal concentration that produces the greatest contrast in K. This optimal concentration is different for each polymer sample and each pore size.

Now, let us make the pore surface attractive to the polymer. The size-exclusion (entropic) effect and the enthalpic effect compete with each other as they do at low concentrations. It is then possible to choose the right surface chemistry to realize a nearly flat MW dependence of K for a polymer of a given chemical composition. Simply put, lines L1 and L2 in Figure 1 compensate for each other to generate the solid line in Figure 1c. Unlike in dilute solutions, a slight change in the interaction does not lead to total exclusion or total adsorption for high-MW components because MW dependence is suppressed. A polymer more attractive to the pore surface than the one for the solid line will follow another nearly flat line, parallel to and slightly above the solid line. As a result, separation of a heterogeneous polymer occurs primarily due to the difference in the polymer components' affinity toward the pore surface. This is the principle of PFC. For a theoretical formulation, based on Flory-Huggins mean-field theory on ternary solutions, see references (*10,29*).

It often happens that the solvent is selective to the polymer and the concentrated solution has large spatial fluctuations in composition. The latter will be visible as haziness. The extreme is a two-phase solution that will look whitish. When this solution is injected onto a column where the stationary phase attracts only one of the components of the polymer, the heterogeneous solution will be segregated between the mobile phase and the stationary phase. This separation mechanism should provide a resolution superior to the one discussed for the solid line Figure 1c.

Unlike separations at low concentrations, the separation at high concentrations does not follow a universal relationship between the retention volume and the chemical composition or MW of the polymer. Rather, efforts have been directed at securing a sufficient amount of polymer in each fraction to allow off-line characterization, often by two or more methods. The latter include infrared absorption (IR) spectroscopy, proton nuclear magnetic resonance (^{1}H-NMR) spectroscopy, differential scanning calorimetry (DSC), and analytical chromatography, as listed in Table 1.

Optimization of Separation

An HOPC/PFC experiment proceeds in the following fashion: First, inject a large volume of a concentrated solution of a given polymer into a column packed with porous particles. Then, collect the eluent into different fractions. Last, characterize the polymer in each fraction using appropriate methods.

Parameters of HOPC/PFC that a user can control include column, solvent, polymer concentration, and flow rate. The flow should be sufficiently slow to permit the transfer of polymer chains in viscous solution between the mobile phase and the stationary phase and to minimize viscous fingering when the solvent enters the solution-filled column. The concentration should be sufficiently high to maintain the condition of semidilute solutions for as many theoretical plates as possible during separation. Below, columns and solvents used in HOPC/PFC are explained in detail. Note that in both HOPC and PFC ternary interactions between polymer, solvent, and surface are present. Therefore, for a given polymer, column and solvent need to be optimized simultaneously.

Columns

Commercial columns are available with different stationary phases. Rigid packing materials such as porous silica are preferred; polymeric gels may collapse irreversibly upon exposure to concentrated solutions. Large particle size (> 15 μm) may be better to decrease high back-pressure.

It is also possible to prepare columns with the user's own packing material. Methods to attach chemical moieties to surface silanol are well known (*30-32*), for example, and a variety of silanization agents are commercially available.

In HOPC the pore surface is modified to prevent adsorption of polymer. Octadecyl (C8) and phenyl substitution are appropriate for separation in an organic medium. Recently it was found that a slightly attractive surface improves the resolution in HOPC (*6,7*). For selection of an appropriate pore size, see reference (*33*).

In PFC a surface that attracts one of the components of the heterogeneous polymer is used. The optimal surface chemistry to separate a given polymer can be found by actually running PFC with various stationary phases. In principle, however, we can use the rule of likes-attract-each-other. For example, in the separation of a random copolymer of styrene and acrylonitrile it is to be expected that a phenyl-substituted surface will prefer the styrene-rich components of the copolymer, whereas a cyanopropyl-substituted surface will prefer the acrylonitrile-rich components. In PFC of ionic polymers in aqueous mobile phase, a stationary phase that bears the opposite charge of the polymer at the pH of separation will give the optimal separation. It often happens for polymers that a specific interaction is present between a pair of repeat units on different molecules. For instance, an oxyethylene unit in poly(ethylene glycol) is attracted to the carboxylic acid group in poly(acrylic acid) to form a complex

(13). As a result of this, we can expect that carboxylic surface moieties will prefer oxyethylene-rich components in the copolymer of ethylene glycol that includes block copolymers. There are other examples of specific interactions that make two polymers miscible in a blend. Previous studies of polymer blends may provide a guide to the selection of appropriate stationary phases for PFC of a given heterogeneous polymer *(34)*.

Solvents

Any solvent that dissolves a given polymer can be used in HOPC. However, it was recently found that a theta solvent for the polymer gives a better resolution compared with a good solvent *(6,7)*.

In general, PFC performs better when it uses a selective solvent that dissolves the surface-repelled components of a given polymer better than the other components. For instance, toluene dissolves the styrene-rich components of poly(styrene-*co*-acrylonitrile) (SAN) better than it does the acrylonitrile-rich components. Therefore, separation of SAN in toluene with a cyanopropyl stationary phase should provide a good resolution.

Examples of Separations

This section describes examples of two-dimensional separations with multiple detection. In the last row of Table I, HOPC–SEC (HOPC in the first dimension, SEC in the second) may be able to uncover components that are hidden in SEC analysis alone. However, HOPC–SEC can be substituted by SEC–SEC, since the amount of polymer needed for SEC analysis in the second dimension is extremely small. Therefore, the method is not explained here.

Chemical Composition Analysis of a Random Copolymer

The first systematic study of PFC was conducted on SAN *(11,14)*. This copolymer consists of polar and nonpolar monomeric units, thus providing a fertile ground to study the effects of surface chemistry, solvent, and polymer concentration on the performance of PFC. The purpose of the studies at that time was to elucidate the separation mechanism and optimize the separation performance rather than to analyze the chemical composition distribution of the copolymer. Only recently was it realized that PFC could analyze the composition distribution. Since the findings of these studies are useful for analytical purposes as well, they are explained here briefly.

Columns (3.9 mm × 300 mm) packed with controlled pore glass (CPG; 497 Å mean pore diameter, 200/400 mesh) whose surface was modified to diphenyl and cyanopropyl were used. An SAN sample with an average weight fraction of styrene of 20% was dissolved in a solvent at 0.134 g/mL and injected at 0.3 mL/min through the pump head of a single-head HPLC pump into the column until the first polymer appeared in the eluent. Then, the injection was switched to the pure solvent, and the eluent was cut into 18 fractions. Later fractions collected more drops. The concentration of polymer in the eluent peaked in fractions 5–8.

The average composition of the copolymer in each of the separated fractions was analyzed using infrared absorption spectroscopy. The styrene unit has a well-defined peak at 1605 cm^{-1}, the acrylonitrile unit at 2239 cm^{-1}. A calibration curve that relates the peak height ratio to the mole fraction ratio was obtained for five samples of SAN with different average compositions. Thus, measurement of the absorption spectrum of each fraction gave x_{ST}, the average mole fraction of styrene in that fraction.

Figure 2 shows x_{ST} as a function of the fraction number. Solutions of SAN in toluene, fluorotoluene, and 1-methyl-2-pyrrolidinone (NMP) were injected into a column packed with cyanopropyl-modified CPG. The span of x_{ST} is the greatest in toluene, followed by fluorotoluene. In contrast, NMP returned a poor separation, and the above trend was reversed. Among the solvents studied toluene is the most selective to styrene-rich components. The selectivity was confirmed by the presence of a so-called heterogeneity mode (35) in dynamic light scattering autocorrelation functions obtained for the same polymer–solvent systems, but it was also visible as solution haziness developed at high concentrations. As expected, the solvent that selectively dissolved the surface-repelled components produced a good separation.

When the stationary phase was diphenyl, dioxane resolved the chemical composition better than NMP and methyl ethyl ketone did (11,14). The early eluent was enriched with acrylonitrile-rich components. Again, the solvent that selectively dissolves the surface-repelled components led to a successful separation. It was also found that the higher the concentration of the injected solution, the better the resolution (11). Thus, the separation mechanism of PFC described in the preceding section was substantiated.

There is also an optimal pore size in PFC. If the pore is too small the polymer is separated by MW. If the pore is too large the surface area is too small

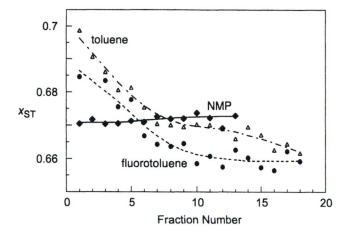

Figure 2. Mole fraction of styrene, x_{ST}, in fractions obtained in PFC of styrene–acrylonitrile copolymer with cyanopropyl stationary phase. Solvents were toluene, fluorotoluene, and NMP. The gray rectangle indicates x_{ST} of the original copolymer. (Reproduced from reference 14. Copyright 2001 Wiley-VCH.)

to effectively interact with the polymer. At the optimal pore size PFC separated the copolymer by a difference in the affinity to the stationary phase. In fact, SEC chromatograms for the fractions obtained were reversed in order (*11*).

In these studies the amount of polymer collected in each fraction was not measured. Said measurement would have allowed us to prepare a plot of the chemical composition distribution of SAN. An example of such plots will be shown for a block copolymer in the last part of this section.

End-Group Analysis of a Telechelic Polymer

Monomethoxy-substituted poly(ethylene glycol) (MePEG) with a structure of $H-(O-CH_2-CH_2)_n-OCH_3$ is widely used as a macroinitiator to grow a second block on the terminal OH and as a precursor to functionalized poly(ethylene glycol) (PEG) (*36*). MePEG is usually prepared by ring-opening polymerization of ethylene oxide with methanol as initiator. However, presence of residual water leads to a diol in the reaction chamber. Since water is more reactive, it is not easy to prepare MePEG that does not contain PEG diol. The chain grows in two directions when initiated with water but in only one direction when methanol starts the polymerization. In the latter case the product will contain a component that is twice as high in MW as the main component and that has diol termini. Presence of diol in an MePEG sample leads to a mixture of diblock and triblock copolymers when another block is grown on the hydroxy ends of PEG. Gelation may occur in some systems containing block copolymer prepared from MePEG because of the triblock component.

Figure 3 shows SEC chromatograms for three commercial samples of MePEG (nominal MW = 5000 g/mol). In addition to the main peak, a second peak at twice as high a MW is evident. A shoulder at MW three times as high and another small peak at MW four times as high are also present. The ratio of the MWs of these monomeric and multimeric components is 1:2:3:4. To identify the terminal groups for each of these components two-dimensional analysis of the MePEG samples was conducted with HOPC in the first dimension and SEC and NMR in the second dimension (8,9).

For HOPC, a column (3.9 mm × 300 mm) was packed with acid-washed controlled pore glass (85 Å mean pore diameter, 100/200 mesh) and filled with water. MePEG was dissolved in water at 50% (w/w) and injected at 0.2 mL/min into the column until the first polymer appeared in the eluent. The eluent was cut into 16 fractions.

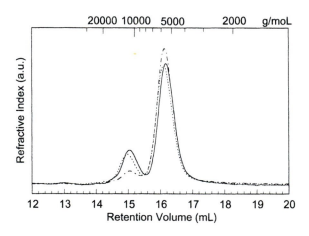

Figure 3. SEC chromatograms for three samples of MePEG. The upper axis indicates MW. (Reproduced from reference 8. Copyright 2002 Elsevier.)

The polymer in the collected fractions was characterized using off-line SEC and ^{1}H-NMR. Figure 4 shows SEC chromatograms for some of the fractions obtained from the MePEG sample shown as a dashed line in Figure 3. A transition from early fractions with large peak areas of multimeric components to late fractions with a prominent monomeric peak is evident. Each chromatogram was fitted with a sum of four normal distribution functions to find the mass fractions of the four components. By assigning terminal chemistry (OH or OCH$_3$) to each component, x_{SEC}, the average number of methoxy units per n oxyethylene units in each fraction, where n units constitute the monomeric component, can be calculated. Since trimeric and tetrameric components are negligible in mass, their end group assignment barely affects x_{SEC}. In the NMR spectrum of each fraction the integral of the methoxy proton peak at 3.24 ppm relative to the integral of the oxyethylene proton peaks at 3.28–3.90 ppm was used to estimate x_{NMR}, the number of methoxy units per oxyethylene unit. The two estimates are related by $x_{NMR} = n\,x_{SEC}$.

The two estimates were obtained for each of early fractions and the original sample. It was found that one methoxy unit per chain in the monomeric component and 0–0.1 methoxy units per chain in the dimeric component give both the optimal proportionality between x_{NMR} and x_{SEC} and the proportionality constant that agrees with the degree of polymerization of the monomeric component. Thus, it was concluded that the monomeric component has monomethoxy–monohydroxy termini whereas the dimeric component has dihydroxy termini. The terminal groups of trimeric and tetrameric components were determined to be diols by NMR analysis of an early eluent in SEC for the early fraction obtained in HOPC. The latter was purified twice to remove the monomeric component. The same conclusion was drawn for the other two samples of MePEG (8).

Composition Analysis of a Block Copolymer

Separation of a block copolymer by PFC produces fractions that have different overall composition, number of blocks, and block lengths. These are related to each other. The rules that govern the separation of random copolymers also apply to the separation of block copolymers. Namely, the resolution is better when the solvent is selective to the copolymer components that are being repelled by the pore surface, and when the polymer concentration is higher.

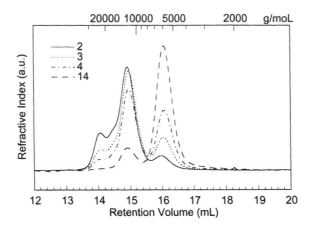

Figure 4. SEC chromatograms for some of the fractions separated from MePEG. The fraction number is indicated in the legend. (Reproduced from reference 8. Copyright 2002 Elsevier.)

Characterization of the separated fractions can be done more easily and accurately for block copolymers as compared to random copolymers: The ^{1}H-NMR spectrum has well-separated, narrow peaks for the protons residing on different blocks of the copolymer. The cleaner baseline of the ^{1}H-NMR spectrum as compared to that of the infrared absorption spectrum provides the necessary advantage. By way of contrast, in random copolymers a variety of microscopic environments caused by sequence and tacticity distributions broaden the resonance peaks to undermine the quantitative capability of NMR analysis. Furthermore, the number of junctions between blocks may also be estimated. When the latter capability is applied to the fractions separated by PFC, it can give information on the distribution in the number of blocks.

The block copolymer selected was PEG–PLLA, where PLLA is poly(L-lactic acid). This copolymer was prepared by ring-opening polymerization of L-lactide with MePEG as a macroinitiator. At early stages of the study (before end-group analysis of MePEG) we were ignorant of the significance of the diol components in the initiator and assumed that the copolymer obtained was purely a diblock (*13,15*).

The length distribution of the PLLA block is broader compared with that of the PEG block. Therefore, a high-MW copolymer tends to have a greater lactate content than does a low-MW copolymer. It is possible to take advantage of this correlation. Separation of the copolymer using porous materials with a pore size sufficiently small to exclude long chains and with a surface that prefers oxyethylene-rich components should have a high resolution, because the HOPC mechanism and the PFC mechanism cooperate. In fact, early studies conducted using simple surface modifications resulted in separations that eluted lactate-rich components in the early fractions (12,13). Among others, a carboxymethyl surface produced the best result in retaining oxyethylene-rich components.

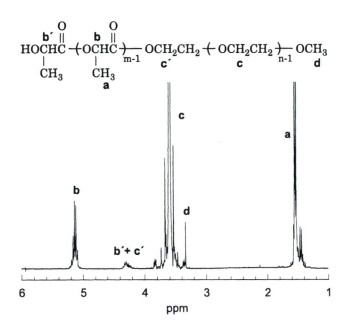

Figure 5. Structure and NMR spectrum of PEG–PLLA block copolymer. Peak assignments are indicated by letters. (Reproduced from reference 15. Copyright 2001 American Chemical Society.)

Reversal of the elution order was accomplished by using porous silica with PLLA brushes (15,16). The latter were grown on surface silanol by ring-opening polymerization of lactide. Figure 6 compares x_{LLA}, the mole fraction of lactate in the eluent, as a function of the cumulative drop count since the detection of the first polymer in the eluent. Separations using carboxymethyl-modified CPG (82 Å pore diameter) and two classes of PLLA-modified CPG (182 and 343 Å) are compared. The solvent was dioxane. Although there is an influence of size-exclusion in early fractions, especially in the separation with the smaller pore size packing, PLLA-modified CPG largely retained lactate-rich components more than it did the other components. The longer PLLA block in the early fractions as compared to the late fractions in the separation by the carboxymethyl stationary phase was confirmed by using HPLC at the critical condition of PEG

260

(*16*), thus masking the PEG block. SEC analysis of each of the fractions obtained in the three separations revealed that the effect of size-exclusion was smaller in the separations with the PLLA columns than it was in the separation with the carboxymethyl-modified column.

To give a flavor of the composition analysis by PFC, Figure 7 shows x_{LLA} as a function of the cumulative mass fraction of the polymer in the eluent. For the carboxymethyl stationary phase, the plot is reversed. As the resolution of PFC is better in the early fractions than it is in the late fractions, the lower left end of the curve for the PLLA stationary phase and the upper right end of the curve for the carboxymethyl stationary phase are considered to be close to reality. Thus, combining the two opposite PFC separations that excel in uncovering different parts of the chemical composition of the polymer we can estimate the overall composition distribution.

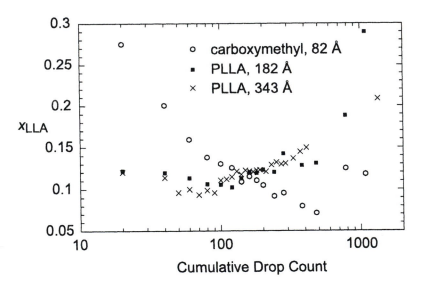

Figure 6. Mole fraction of lactate, x_{LLA}, plotted as a function of the cumulative drop count of the eluent after detection of the first polymer. Results are shown for the three stationary phases indicated in the legend. (Reproduced from reference 16. Copyright 2002 Elsevier.)

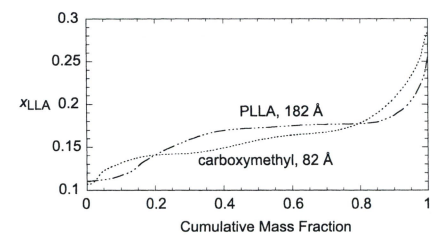

Figure 7. Mole fraction of lactate, x_{LLA}, plotted as a function of the cumulative mass fraction of the polymer in the eluent. Results are shown for two columns. (Reproduced from reference 15. Copyright 2001 American Chemical Society.)

Concluding Remarks

Two-dimensional analysis of complex polymers with HOPC or PFC in the first dimension was reviewed. Unlike other chromatographic methods, the preparative methods inject a large volume of a concentrated solution of the polymer into an analytical-size column. The first dimension is consumed to secure a large amount of polymer in each of the fractions collected, disregarding the relationship between the retention time and the polymer property. The large amount allows various off-line characterizations of each fraction in the second dimension, often by more than one methods. Although the procedure is tedious, the analysis provides detailed information of polymer composition. Examples were shown for chemical composition analysis of random copolymers, end-group analysis of telechelic polymers, and composition analysis of block copolymers. To facilitate application to other complex polymers, the separation principles and the guidelines for selection of a column and a mobile phase were also described.

References

1. Luo, M.; Teraoka, I. *Macromolecules* **1996**, *29*, 4226.
2. Luo, M.; Teraoka, I. *Polymer*, **1998**, *39*, 891.
3. Xu, Y.; Teraoka, I. Senak, L.; Wu, C.-S. *Polymer*, **1999**, *40*, 7359.
4. Matsuyama, S.; Nakahara, H.; Takeuchi, K.; Nagahata, R.; Kinugasa, S.; Teraoka, I. *Polym. J.* **2000**, *32*, 249.
5. Kagawa, N. *Tosoh Res. Tech. Rep. (in Japanese)* **2000**, *44*, 15.
6. Lee, D.; Gong, Y.; Teraoka, I. *Macromolecules* **2002**, *35*, 7093.
7. Lee, D.; Teraoka, I. *J. Chrom. A.* **2003**, *996*, 71.
8. Lee, D.; Teraoka, I. *Polymer* **2002**, *43*, 2691.
9. Lee, D.; Teraoka, I. *Biomaterials* **2003**, *24*, 329.
10. Luo, M.; Teraoka, I. PMSE Preprints **1996**, *75*, 299.
11. Xu, Y.; Teraoka, I. *Macromolecules* **1998**, *31*, 4143.
12. Fujiwara, T.; Kimura, Y.; Teraoka, I. *J. Polym. Sci. Polym. Chem. Ed.* **2000**, *38*, 2405.
13. Fujiwara, T.; Kimura, Y.; Teraoka, I. *Polymer* **2001**, *42*, 1067.
14. Zheng, H.; Teraoka, I.; Berek, D. *Macromol. Chem. Phys.* **2001**, *202*, 765.
15. Lee, D.; Teraoka, I.; Fujiwara, T.; Kimura, Y. *Macromolecules* **2001**, *34*, 4949.
16. Lee, D.; Teraoka, I.; Fujiwara, T.; Kimura, Y. *J. Chrom. A.* **2002**, *966*, 41.
17. Guttman, C. M.; DiMarzio, E. A.; Douglas, J. F. *Macromolecules* **1996**, *29*, 5723.
18. Berek, D. *Prog. Polym. Sci.* **2000**, *25*, 873.
19. Teramachi, S.; Hasegawa, A.; Shima, Y.; Akatsuka, M.; Nakajima, M. *Macromolecules* **1996**, *29*, 992.
20. Glöckner, G. *Gradient HPLC of Copolymers and Chromatographic Cross-Fractionation*, Springer, Berlin, 1991.
21. Chang, T.; Lee, H. C.; Park, W.; Ko, C. *Macromol. Chem. Phys.* **1999**, *200*, 2188.
22. Teraoka, I. *Polymer Solutions: An Introduction to Physical Properties*, John Wiley, New York, 2002.
23. Daoud, M.; de Gennes, P. G. *J. Phys. (Paris)* **1977**, *38*, 85.
24. Teraoka, I.; Langley, K. H.; Karasz, F. E. *Macromolecules* **1993**, *26*, 287.
25. Wang, Y.; Teraoka, I. *Macromolecules* **1997**, *30*, 8473.
26. Wang, Y.; Teraoka, I. *J. Chem. Phys.* **2001**, *115*, 1105.
27. Teraoka, I.; Zhou, Z.; Langley, K. H.; Karasz, F. E. *Macromolecules* **1993**, *26*, 3223
28. Teraoka, I.; Zhou, Z.; Langley, K. H.; Karasz, F. E. *Macromolecules* **1993**, *26*,, 6081.
29. Xu, Y. Ph.D. Thesis, **1999**, Polytechnic University.

30. Arkles, B. *Chemtech* **1977**, *7*, 766. Also at http://www.gelest.com/Library/08Tailor.pdf.
31. Plueddemann, E. P. *Silane Coupling Agents*, 2nd Ed. Plenum, New York, 1991.
32. Neue, U. D. *HPLC Columns: Theory, Technology, and Practice*, Wiley-VCH, New York, 1997.
33. Teraoka, I. in *Column Handbook for Size Exclusion Chromatography*, Wu, C.-S. ed. Academic Press, San Diego, 1999.
34. Brandrup, J.; Immergut, E. H.; Grulke, E. A. *Polymer Handbook*, 4th Ed. John Wiley, New York, 1999.
35. Koňák, Č.; Fleischer, G. *Macromolecules* **1997**, *30*, 1457.
36. http://www.nektar.com/pdf/shearwater_catalog.pdf.

Dynamic Surface-Tension Detection, Data Reduction, and Band Broadening

Chapter 16

Size-Exclusion Chromatography with Dynamic Surface Tension Detection: Analysis of Polymers and Proteins

Robert E. Synovec[1], Bethany A. Staggemeier[1], Emilia Bramanti[2], Wes W. C. Quigley[1], and Bryan J. Prazen[1]

[1] Department of Chemistry, Center for Process Analytical Chemistry, University of Washington, Box 351700, Seattle, WA 98195–1700
[2] Istituto per i Processi Chimico-Fisici, Laboratory of Instrumental Analytical Chemistry, Consiglio Nazionale delle Ricerche-CNR, Via G. Moruzzi 1, 56124 Pisa, Italy

Multidimensional analysis of linear and branched polymers and proteins by size exclusion chromatography with dynamic surface tension detection is presented. In particular, the application of this hyphenated chromatographic technique is applied to the analysis and characterization of poly (ethylene) glycol (PEG) polydispersity, molecular sizing of branched PEGs, and separation and analysis of protein samples. Benefits of multidimensional chromatographic detection for SEC and selective surface activity analysis are briefly discussed. The dynamic nature of the surface tension signal is shown to provide additional chemical information and selectivity.

Characterization of molecules based on size is a continuing subject of extensive research. Size exclusion chromatography (SEC) is one common analytical technique used to determine the size of molecules (1-7). When SEC is coupled with a detector that provides multidimensional information, the resulting separation provides increased analysis and characterization capabilities with little or no increase in analysis time (8). SEC is commonly used for the separation and analysis of both polymers and proteins, many of which are also surface-active species. These are often used in industrial practices and products because of their ability to concentrate and align at interfaces to lower surface tension (9-11). Utilizing an inherently selective physical property such as surface activity in the detection of eluting analytes allows for an added dimension of information to be gained regarding the sample during the SEC analysis. Furthermore, multidimensional characterization of analytes or mixtures can then be obtained from a single study. Surface tension is an emerging form of multidimensional detection that should enhance the utility of SEC. Herein, we explore the combination of SEC with a dynamic surface tension detector (DSTD).

Background/Theory

The DSTD is a drop-based analyzer that, when used as a chromatographic detector, provides real-time dynamic surface pressure measurements of components as they elute. Because this is a flow-based detector operating at flow rates similar to analytical chromatography systems, it is well suited for coupling with SEC analysis. The DSTD provides an added dimension of sensitivity and selectivity based on a time-dependent kinetic surface activity signal that is related to the size and complexity of the surface-active analytes or systems. This results in multidimensional data that provides added characterization information about a sample. The DSTD provides data on eluting components as a measure of the change in surface tension of a growing drop due to the presence of a surface-active analyte. The DSTD has been shown previously to be a sensitive and selective detector for a variety of chromatographic applications (12-14). Here we discuss the utility of SEC-DSTD as a multidimensional hyphenated chromatographic technique for the analysis of polymers and proteins. Specifically in this chapter we will present examples of SEC-DSTD data of the analysis and characterization of water-soluble, surface-active linear and branched polymers, and proteins.

The DSTD enables the measurement of the changing surface pressure of a flowing liquid at the air/liquid interface of growing and detaching drops. The SEC-DSTD comprises a pump, an injector, a pressure transducer, and a capillary sensing tip with connections made of poly (ether ether) ketone (PEEK) tubing and controlled by a computer (Figure 1). The SEC column is inserted between the injection valve and the capillary sensing tip for detection and analysis of a separated surface-active sample. A nozzle is situated to deliver an air burst at a ~45° angle from the sensing tip at regularly programmed intervals. This reproducible pneumatic detachment of drops allows high data density collection suitable for chromatographic detection. The interested reader can find further details on the DSTD instrumentation in previous reports (12-17).

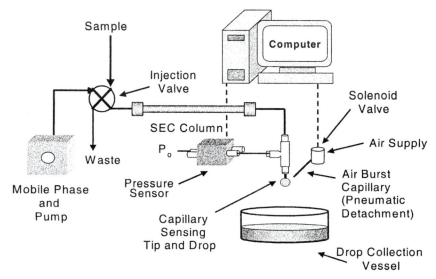

Figure 1. Schematic of the dynamic surface tension detector.

Drop-based analysis methods have been shown to be sensitive detection techniques with an array of instrumentation designs and applications (18). Thus, there are a variety of pressure sensor designs that would work well with the DSTD, such as capacitance, piezoelectric, and solid-state pressure sensors (19-21). The DSTD, as currently configured, utilizes a simple design wherein measurements are recorded based on the response of a membrane–based pressure sensor. The sensor is divided into two cavities by the flexible membrane as shown in Figure 2. One side of the membrane is connected via tubing to the capillary sensing tip, where drops are forming, while the other is open to the atmosphere. Thus, the sensor provides a measure of the differential pressure between the interior of the growing drop in relation to the atmospheric pressure P_o.

In Figure 2A, mobile phase (eg., water) flows through the tubing forming drops at the capillary tip. As these drops form, the surface tension of the growing drop results in pressure on the sensor membrane, which is open to atmosphere. This causes displacement of the membrane, which is recorded as drop pressure. Similarly, in Figure 2B, mobile phase containing surface-active analyte forms drops at the capillary sensing tip. However, in the case of drops formed in the presence of surface-active analyte, the surface tension of the growing drop is lowered, thus the displacement of the sensor membrane is less, as can be seen in the figure, and thus, the recorded drop pressure is less.

Each drop formed at the capillary sensing tip can be considered as an individual sample from which pressure data is obtained throughout drop growth resulting in a 'drop profile'. In the current instrument configuration to measure the air/liquid interface, the drops are reproducibly detached pneumatically by a burst of air at low volumes (1-6 μL) (15). When drops are detached at sufficiently low volume, elongation due to gravity is not an issue as drop growth will be effectively spherical. This ensures that the drop radius function r(t) is the same from one drop to the next. The surface tension at the air/liquid interface is related to the time-dependent modified Young-Laplace equation as follows (16, 17):

$$P(t) = 2 \, \gamma(t) \, / \, r(t) + P_C \qquad (1)$$

where P(t) is the differential pressure across the drop interface throughout drop growth, relative to atmospheric pressure, γ(t) is the surface tension at the interface, and r(t) is the radius of the drop as a function of time during drop growth. P_C represents the offset pressure and viscous losses in the tubing, and is generally independent of time in most SEC applications.

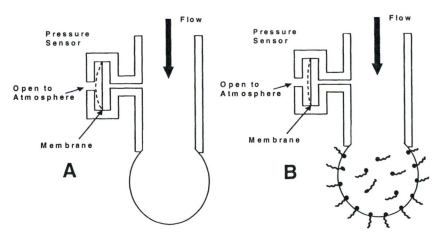

Figure 2. Diagram of pressure sensor and capillary tip during drop growth. (A) Displacement of the sensor membrane during formation of a mobile phase droplet. (B) Displacement of the sensor membrane during formation of a droplet containing surface-active analyte.

Figure 3 shows the raw pressure signal, P(t) defined in equation 1, from an eluting, surface-active component as it might elute from a SEC column. As the surface-active analyte passes through the capillary sensing tip, the condition shown in Figure 2B, the resulting drop pressure decreases. The pressure sensor provides a data vector for each drop, which is related to the dynamic surface tension of the drop as in equation 1. A clear example of the data vectors, termed drop profiles, is shown in the bottom of Figure 3. Since the drop profile is a function of time, kinetic information on surface tension lowering is obtained from each drop profile. The kinetic information is also a function of the growing drop radius, and evaluation of the kinetic characteristics of samples requires a more detailed consideration of drop growth, as is presented in recent manuscripts (12, 15, 16).

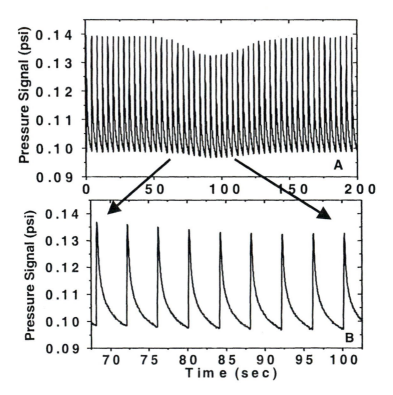

Figure 3. Typical raw pressure sensor data. (A) Elution of a surface-active analyte. (B) Enlargement of the front edge of the eluting surface-active analyte.

From this time-dependent pressure response, given by equation 1, and the definition of the dynamic surface pressure as the surface tension of the mobile phase minus the surface tension of either the analyte or standard in the mobile phase, the following condensed equation has been derived to readily calibrate the eluting analyte signal to a known standard surface pressure, requiring the collection of three drop profiles: those of mobile phase ($P(t)_M$), the analyte ($P(t)_A$), and the standard ($P(t)_S$) (12, 15):

$$\pi(t)_A = \pi_S \left[(P(t)_M - P(t)_A) / (P(t)_M - P(t)_S) \right] \qquad (2)$$

where π_S is the known surface pressure of a standard component, selected to have a surface pressure independent of time, and $\pi(t)_A$ is the experimentally obtained surface pressure of the analyte. Thus, the pressure offset P_C is eliminated in the surface pressure calculations, and the drop radius function $r(t)$ in equation 1 is eliminated through the ratio of the surface pressures, $\pi(t)_A/\pi_S$, thus the sample surface tension information remains. A positive $\pi(t)_A$ corresponds to a decrease in surface tension, $\gamma(t)_A$.

The dynamic surface pressure obtained from equation 2 readily allows visualization, explanation, and comparison of the surface activity of eluting analytes. The surface pressure profile for an analyte that diffuses to and orients at the drop surface quickly is seen to be nearly constant throughout the two-second drop growth. In contrast, a more complex system or larger molecule may begin drop growth with little effect on the surface tension of the drop; however, throughout drop growth, the analyte is able to diffuse to the interface and orient at the surface so that toward the end of drop growth it has produced a significantly higher surface pressure. This dynamic profile would not be obtained from a static surface activity measurement, as only the overall maximum surface pressure would be obtained.

The on-line calibration procedure (equation 2) is illustrated in two examples in Figure 4. Typical time-dependent pressure data, P(t), for applying the calibration procedure are shown in Figures 4A and 4C, where the mobile phase is water and the standard is 5% acetic acid in water. Figure 4B shows the dynamic surface pressure of 1.0 mM sodium dodecyl sulfate (SDS), a non-kinetically hindered sample, where the dynamic surface pressure is nearly constant throughout drop growth. Again, note that as defined, an increase in surface pressure indicates a decrease in sample surface tension, and thus, is indicative of an increase in surface activity at the air/liquid interface. In Figure 4D the dynamic surface pressure plot for a typical kinetically hindered sample is shown. Inspection of Figure 4D reveals that the sample appears to have a signal, i.e., surface activity, near that of water early in drop growth, and then behaves with much more surface activity later in drop growth, much like that of the standard. During a SEC separation, a surface pressure plot is obtained for every two-second drop during a sample run. The data are readily arranged as a three-dimensional surface pressure plot where the x and y axes are elution run time and drop time, and the z-axis is surface pressure.

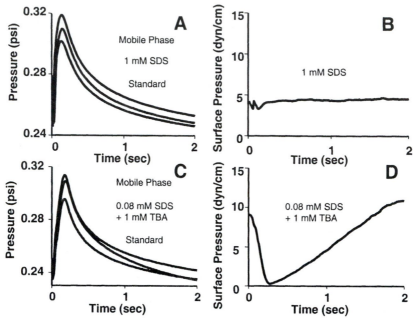

Figure 4. Raw data and surface pressure plots of a non-kinetically hindered analyte and a kinetically hindered analyte. (A) Overlays of individual drop profiles, P(t), (middle) 1 mM SDS sample, (lower) standard (5% acetic acid), and (upper) water (mobile phase) that have been extracted from the raw pressure signals as shown in Figure 3A. (B) Surface pressure plot for SDS resulting from calibration via equation 2. (C) Overlays of individual drop profiles, P(t), (middle) mixture of 0.08 mM SDS and 1.0 mM non-surface-active tetrabutyl ammonium salt (TBA), (lower) standard (5% acetic acid), and (upper) water (mobile phase) that have been extracted from the raw pressure signals. (D) Surface pressure plot for the surface-active, but kinetically hindered, SDS / TBA mixture resulting from calibration via equation 2.

SEC-DSTD of Polymers

Miller, et. al. have shown the selectivity of the DSTD, when coupled with SEC separations, allows the determination of poly(ethylene glycol) (PEG) polydispersity (12). Figure 5A shows the three-dimensional plot of a SEC-DSTD separation of two different molecular mass PEGs wherein the dynamic signal shows that the larger molecular mass species requires significantly more time to adsorb and arrange at the surface of each forming drop. The two species are clearly distinguishable by using the DSTD, as the slope of the rise in surface pressure throughout drop growth for the earlier-eluting, larger PEG increases during each two-second drop growth. The surface pressure signal is a function of analyte concentration. However, the dynamic range is rather limited and care must be taken to apply the DSTD in the near-linear portion of the calibration curve for analytes of interest. This same data is shown as a contour plot in Figure 5B, where the white background indicates the baseline signal and the gray-scale image indicates signal with the darkest regions of the plot representing highest signal. The utility of obtaining this type of selective multidimensional data is shown clearly in Figure 6.

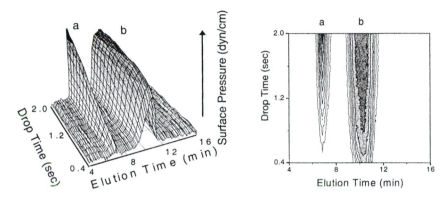

Figure 5. SEC-DSTD separation of two different molar mass PEGs, PEG 22,000 and PEG 1470. (A) Three-dimensional surface pressure plot of the SEC-DSTD separation. (B) Surface pressure contour plot of the SEC-DSTD separation. (Reproduced from reference 6. Copyright 2000 ACS.)

In Figure 6A, refractive index traces of two polymer mixtures with a polydispersity index (Mw/Mn) of 1.58 and 1.65 separated by SEC are indistinguishable from each other by refractive index detection. However, because the dynamic surface activity signal is obtained from the rearrangement of molecules at the surface of the drop, the difference of the DSTD signals obtained from the same separations yields three distinct bands due to the three components present in the polymer mixture (Figure 6B). The polydispersity of PEG samples was quantified using SEC-DSTD and partial least squares (PLS) data analysis yielding a relative precision of ~ 1%, which corresponds to being able to distinguish differences in polydispersity of ~ 0.02 (12).

A second example illustrating the added selectivity of coupling the DSTD with SEC can be seen in the following characterization study of branched poly(ethylene glycols). In this study, three PEGs all with a molecular mass of 20,000 g/mol and varied degrees of branching were obtained for analysis. The basic structures of these polymers are illustrated in Figure 7, where the 3-armed

PEG has a larger molecular size than a 4-armed PEG, which has a larger molecular size than the 8-armed PEG.

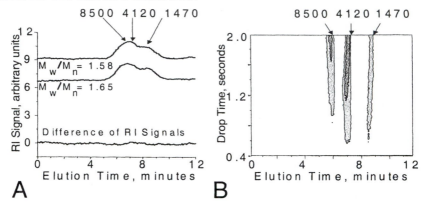

A
B

Figure 6. SEC-RI versus SEC-DSTD separation of PEG mixtures.
(A) Indistinguishable RI traces of 3-component PEG mixtures of different
polydispersity comprising PEG 8,500, 4,120, and 1,470. (B) Contour plot
showing the difference of DSTD signals for the samples shown in Figure 6A,
analyzed by SEC-DSTD. (Reproduced from reference 6. Copyright 2000 ACS.)

SEC analysis of a series of standard linear PEGs of varied molecular mass and these three, branched polymers was performed with detection by the DSTD. The resulting calibration curve is shown in Figure 8. Although the branched polymers each have a true molecular mass of 20,000 g/mol, their molecular size differs due to the branching shown in Figure 7. This results in an apparent molecular mass determined by SEC elution that is smaller than the actual mass for the branched polymers.

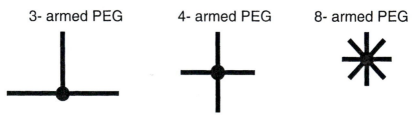

3- armed PEG   4- armed PEG   8- armed PEG

Figure 7. Illustration of the relative sizes of branched 20,000 g/mol PEGs

In Figure 9, data from flow injection experiments of the same branched polymers with DSTD analysis shows the utility of the kinetic signal obtained with this detector. The DSTD easily distinguishes the 3 polymers (see Figure 9B), since the selectivity is based on the dynamic nature of the DSTD response.

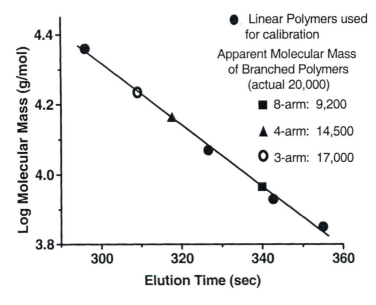

Figure 8. Calibration of linear PEGs analyzed by SEC-DSTD showing the apparent molecular mass of the branched polymers based on SEC elution time.

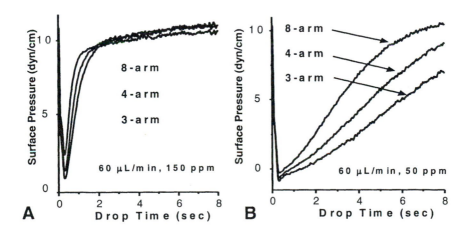

Figure 9. Resulting dynamic surface pressure plots from the flow injection - DSTD analysis of the branched poly(ethylene) glycols at two concentrations showing size selectivity both analogous and complementary to SEC. (A) 8-arm, 4-arm, and 3-arm branched 20,000 g/mol PEG each at 150 ppm. (B) 8-arm, 4-arm, and 3-arm branched 20,000 g/mol PEG each at 50 ppm.

On the other hand, static surface tension measurement techniques often applied would yield no selectivity, as indicated in the 150 ppm plot, since a static measurement only provides the surface pressure at the end of the drop growth where each sample has achieved the same surface pressure. With regard to molar mass of the branched PEGs, the kinetic hindrance observed in the DSTD plots (Figure 9) is inversely correlated to the SEC elution time (Figure 8). Because the dynamic signal relates to the required diffusion and adsorption time of surface-active molecules (where larger molecules tend to require longer time to reach maximum surface activity), the dynamic information in the DSTD signal is complementary to sizing information gained through SEC.

SEC-DSTD of Proteins

The DSTD has been utilized as a selective characterization and detection tool for proteins, some of which are surface-active in their native state, and many of which have also been shown to become highly surface active under conditions of denaturation (11, 13, 14, 22). In the following examples, SEC-DSTD was performed in buffer solutions with post-column protein denaturation by guanidine thiocyanate prior to the DSTD. Figures 10 A and B show results of a SEC-DSTD separation of a three-protein standard mixture: bovine serum albumin (BSA, 65,000 g/mol), myoglobin (Myo 16,125 g/mol) and cytochrome c (Cyt C, 12,327 g/mol). The inherent utility and selectivity of dynamic surface tension detection for complex protein mixtures can be seen readily in Figures 10B and 10C where shortening the DSTD drop time from 4 to ~ 2 seconds would allow for selective detection of only the Cyt C protein. Note that a modified calibration procedure was employed for the SEC-DSTD of proteins utilizing four drop profiles (14, 22). This calibration procedure is an important step in the evolution of the DSTD; it allows samples to be utilized in many solvent matrices, i.e., mobile phases such as guanidine thiocyanate, without changing the standardization step of the calibration, utilizing a standard in a water mobile phase (with a readily obtained surface pressure). Gravity did not adversely affect the drop shape with respect to the calibration procedure. Similarly, Figure 11 shows the results of a SEC-DSTD analysis of a second three-protein standard mixture containing β-lactoglobulin (BLG, dimer 36,000 g/mol), carbonic anhydrase (CA, 31,000 g/mol), and α-chymotrypsinogen (α-Chy, 25,700 g/mol). The proteins are again readily visually separated based upon the multidimensional data obtained by SEC-DSTD.

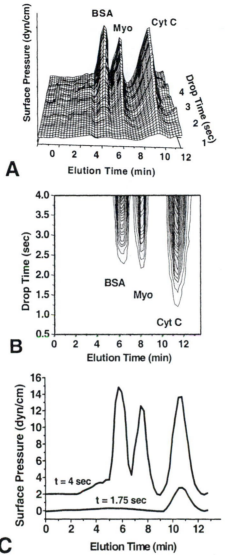

Figure 10. SEC-DSTD separation of three commercial standard proteins with post-column protein denaturation: (BSA (1500 mg/L), Myo (1000 mg/L), Cyt C (1800 mg/L)). (A) Three-dimensional surface pressure plot of BSA, Myo, and Cyt C. (B) Contour surface pressure plot of BSA, Myo, and Cyt C. (C) Plot of elution time vs. surface pressure for the same separation of BSA, Myo and Cyt C taken at the end of drop time (4 sec) and at 1.75 seconds (part-way through drop growth). The plots have been offset for clarity.

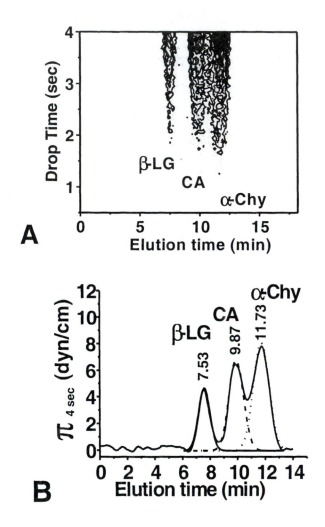

Figure 11. (A) Contour surface pressure plot of a SEC-DSTD separation of the three-protein mixture containing β–LG, CA, and α-Chy (β–LG (1000 mg/L), Myo (1000 mg/L), α-Chy (1500 mg/L)). (B) Results of the same SEC-DSTD separation shown in A, plotted as mean surface pressure at drop detachment ($\pi_{4\,sec}$) vs. elution time.

Conclusions

The DSTD can be employed as a valuable detection method for multidimensional analysis of SEC separations. The DSTD provides a simple, yet dynamic surface tension-based measurement without the need for complicated optical measurements. The DSTD is routinely able to detect analytes at the low ppm level (18), and techniques to enhance the obtainable signal of some analytes have also been successfully employed (22-24). As a drop-based detector for surface activity, the precision of DSTD measurements is similar to that of many commercial instruments for measuring static surface tension with $3\sigma \sim 0.2$ dyn/cm (16, 25). It should be noted that the current DSTD instrumentation is optimized to take advantage of the kinetic analyte signal during drop growth. The limit of detection (LOD) for proteins denatured in guanidine thiocyanate ranges from 5 to 14 ppm (22). In comparison, the LOD for SDS in the presence of cations such as TBA and Cr^{3+} is ~ 0.3 ppm (23). Selecting a more flexible sensing membrane might enhance these detection limits even further.

As a stand-alone analyzer, the DSTD is used to obtain size and structure information for analytes with inherent surface activity. When hyphenated with SEC, the resulting SEC-DSTD technique should enhance the utility of the DSTD as a selective detector. Examples of SEC-DSTD analysis of branched and linear polymer samples have been presented. In addition, SEC-DSTD analyses of protein samples have also been achieved. Interpretation of the complementary dynamic DSTD signal for both molecular sizing and selective detection has been discussed. The hyphenated technique SEC-DSTD has been shown to be a useful multidimensional tool for complex chromatographic analysis of surface-active analytes, increasing the characterization information obtained during analysis. Implementing SEC-DSTD with non-aqueous separations is also being explored, and this should broaden the scope of the technique. In this case, the DSTD would be employed in a liquid/liquid drop-based interface mode, forming non-aqueous solvent drops in an immiscible solvent such as water.

References

(1) Smith, P. B.; Pasztor, A. J.; McKelvy, M. L.; Meunier, D. M.; Froelicher, S. W.; Wang, F. C. Y. *Anal. Chem.* **1999,** *71*, 61R-80R.
(2) Barth, H. G. In *Handbook of HPLC*; Katz, E., Ed.; Marcel Dekker: New York, 1998; Vol. 78, pp 273-292.
(3) Barth, H. G.; Boyes, B. E.; Jackson, C. *Anal. Chem.* **1996,** *68*, 445R-466R.
(4) Barth, H. G.; Boyes, B. E.; Jackson, C. *Anal. Chem.* **1998,** *70*, 251R-278R.
(5) Wen, J.; Arakawa, T.; Philo, J. S. *Anal. Biochem.* **1996,** *240*, 155-166.

280

(6) le Maire, M.; Ghazi, A.; Moller, J. In *Strategies in Size Exclusion Chromatography*; Potschka, M., Dubin, P., Eds.; American Chemical Socity: Washington, D.C., 1996, pp 36-51.

(7) Hunt, B.; Holding, S. *Size Exclusion Chromatography*; Chapman and Hall: New York, 1989.

(8) Pasch, H. *Adv. in Polymer Sci.* **2000,** *150*, 1-66.

(9) Hunt, B. J.; James, M. I. *Polymer Characterization*; Chapman and Hall: Glasgow, 1993.

(10) Tsujii, K. *Surface Activity: Principle, Phenomena and Applications*; Academic Press: San Diego, 1998.

(11) Magdassi, S. In *Surface Activity of Proteins*; Magdassi, S., Ed.; Marcel Dekker, Inc: New York, 1996.

(12) Miller, K. E.; Bramanti, E.; Prazen, B. J.; Prezhdo, M.; Skogerboe, K. J.; Synovec, R. E. *Anal. Chem.* **2000,** *72*, 4372-4380.

(13) Olson, N. A.; Skogerboe, K. J.; Synovec, R. E. *J. Chromatogr. A* **1998,** *806*, 239-250.

(14) Bramanti, E.; Quigley, W. W. C.; Sortino, C.; Beni, F.; Onor, M.; Raspi, G.; Synovec, R. E. *J. Chromatogr. A* **2003,** *Accepted.*

(15) Miller, K. E.; Synovec, R. E. *Anal. Chim. Acta* **2000,** *412*, 149-160.

(16) Miller, K. E.; Skogerboe, K. J.; Synovec, R. E. *Talanta* **1999,** *50*, 1045-1056.

(17) MacLeod, C. A.; Radke, C. J. *J. Coll. Interf. Sci.* **1993,** *160*, 435-448.

(18) Miller, K. E.; Synovec, R. E. *Talanta* **2000,** *51*, 921-933.

(19) van der Heyden, F.; Blom, M.; Gardeniers, J.; Chmela, E.; Elwenspoek, M.; Tijssen, R.; van der Berg, A. *Sensors and Actuators B* **2003,** *92*, 102-109.

(20) Luck, R.; Agba, E. I. *ISA Transactions* **1998,** *37*, 65-72.

(21) Fleury, M.; Smullin, L. D. *J. Vac. Sci. Technol. A* **1995,** *13*, 3003.

(22) Quigley, W. W. C.; Bramanti, E.; Staggemeier, B. A.; Miller, K. E.; Nabi, A.; Skogerboe, K. J.; Synovec, R. E. *Anal. Bioanal. Chem.* **2003,** *Accepted.*

(23) Quigley, W. W.; Nabi, A.; J, P. B.; Lenghor, N.; Grudpan , K.; Synovec, R. E. *Talanta* **2001,** *55*, 551-560.

(24) Young, T. E.; Synovec, R. E. *Talanta* **1996,** *43*, 889-899.

(25) Quigley, W. W. C. Doctoral Dissertation, University of Washington, Seattle, 2002.

Chapter 17

Data Reduction in Size-Exclusion Chromatography with Molar Mass Sensitive Detectors

Yefim Brun

DuPont Central Research & Development, Experimental Station,
Wilmington, DE 19880–0228

A comprehensive analysis of recent improvements in data reduction in triple-detector SEC is presented. Using direct measurements of a sample's hydrodynamic volume across the entire polymer distribution, different sources of non-ideality of size-exclusion separation can be analyzed as well as their effect on accuracy of the calculated molar mass distribution and polymer solution properties. Elimination of concentration detector traces may significantly improve the quantitation of polymers with long high-molar-mass tails, e.g., caused by long-chain branches. Novel approaches to model fitting of calibration and structural curves describing the relations between polymers' structural parameters are discussed in detail. The characterization of an ethylene-methyl acrylate copolymer illustrates the quantitative approaches.

The introduction of molar mass-sensitive detectors, such as the multi-angle laser light scattering photometer and the capillary viscometer, as on-line detectors in an SEC set-up allows for eliminating or at least for significantly reducing the effect of polydispersity in studies of dilute polymer solution properties. Besides an accurate molar mass distribution (MMD), a single chromatographic run provides information about these properties across the entire polymer distribution, i.e., for near-monodisperse polymer fractions with different size of macromolecules. Such capabilities may play a crucial role in elucidating the configuration and conformational statistics of macromolecules. For example, the effects of branching on size of macromolecules and other properties of materials in solution can be increasingly counter-balanced, and in some cases even fully masked, by an increasing polydispersity (*1,2*). In spite of numerous obvious successes (*3*) in accurate calculation of the MMD of polymers with complex architecture (one such example includes a validation of percolation and mean-field theories in describing the MMD of randomly branched polymers (*4,5*)) and their global properties (branching contraction factors, fractal dimension, etc.) it is generally agreed that the power of this analytical technique is at present not fully utilized (*2*). It is a strong perception that due to poorly understood uncertainties the interpretation of elution chromatograms in terms of polymer properties in many cases is extremely difficult and ambiguous (*2*). In other words, data reduction often becomes a bottleneck of an otherwise very promising technique.

Let us briefly discuss complications which can potentially compromise the accuracy of polymer characterization by multidetection SEC. Some of the complications come from the possible non-ideality of the separation itself and are not directly related to multidetection. Thus, significant shear degradation of high molar mass polymers can be caused by an overly high linear velocity inside the columns. Non-steric interaction between a solute and a stationary phase may also occur during separation. Band broadening produces local polydispersity, i.e. polydispersity within each slice of a separation profile. Elimination of the column calibration with external standards substantially reduces the adverse effect of the last two types of non-ideality. However, portions of the polymer may sometimes be irreversibly adsorbed onto the separation gel, and important information about polymers can be lost due to significant band broading even when specially designed mathematical tools are applied for axial dispersion correction (*3*). Therefore, selection of adequate stationary and mobile phases and of other chromatographic conditions (e.g., flow rate) affecting the efficiency of the separation remain important steps in the SEC method development, even with multidetection.

Local polydispersity can also be caused by different types of structural or compositional heterogeneity of macromolecules in complex polymers, such as branched polymers, copolymers, or polymer blends. Proper estimation of this polydispersity and its effect on the measured MMD and polymer solution properties seem imperative for adequate polymer characterization (*6,7*).

Several additional initial parameters are necessary (measured preferably by independent experiments) in multi-detector SEC as compared to its conventional (single-detector) counterpart. These include sample-independent system parameters such as instrument calibration constants, flow rate, interdetector volumes, and injection volume, as well as the concentration of sample injected into a chromatographic system, the specific concentration detector's increment (e.g., refractive index increment), and even polymer-polymer interaction parameters if such interaction in polymer solution cannot be completely eliminated by reducing the polymer concentration (e.g., for very large macromolecules). Sometimes even small changes in these parameters can produce significant differences in calculated polymer solution properties, so that accurate determination of the former is an absolute prerequisite.

An even more challenging problem is the effect of the detectors' sensitivities: molar mass-sensitive detectors have usually low response at the low molar mass end of a polymer's MMD, and concentration detectors have low response at its high end. Consequently, the local (slice) values of molar mass and intrinsic viscosity, calculated as ratios of the corresponding detectors' responses, show high noise levels at both ends of the polymer distribution making related experimental points practically useless. However, these data could be very important, for example in the determination of viscoelastic properties of polymer melts, such as dynamic moduli and steady-state recovery compliance, which are extremely sensitive to minute variations in the high molar mass tail of a polymer's MMD (8,9).

The proper choice of data analysis methodology is the key to resolving the aforementioned problems in SEC with molar mass-sensitive detectors. In this paper we demonstrate several recent advances in such methodology. First, we show how multidetection helps in troubleshooting various non-ideal (non-size-exclusion) effects, as well as the other contributors to local polydispersity. This is achieved by comparing the measured hydrodynamic volume of macromolecules with the hydrodynamic volume corresponding to an ideal size-exclusion separation. The other advances include elimination of a concentration detector in quantitating the high molar mass portion of a polymer's molar mass distribution, as well as a new approach to model-fitting of the calibration and structural curves. In the latter approach a complete chromatographic profile obtained from one detector is compared, in a least-squares sense, to a model that is a function of the responses from the other detector(s). Such approach has two benefits: it allows the inclusion of regions in the least-squares fit that contain very low, close to baseline, signals from one or more detectors without data extrapolation or elimination, and it also provides a direct tool to verify different theoretical models describing polymer solution properties as well as permitting calculation of the parameters of these models.

Dilute Solution Properties and Multi-Detector SEC

General Equations

When a differential refractometer, capillary viscometer, and low-angle laser light scattering photometer are all coupled to a chromatographic system, concentration C_i, molar mass M_i, and intrinsic viscosity $[\eta]_i$ of each slice i, associated with a particular elution volume V_i, can be calculated by solving the following well-known equations for the measured slice values of the refractive index change due to a polymer in solution, ΔN_i, specific viscosity, $\eta_{sp,i}$ and excess Rayleigh ratio at zero-scattering angle, $R_i(0)$:

$$\Delta N_i = (dn/dc)_i \, C_i \tag{1}$$

$$\eta_{sp,i} / C_i = [\eta]_i (1 + k_{H,i} \, C_i \, [\eta]_i) \tag{2}$$

$$K_{LS} (dn/dc)_i^2 \, C_i \, M_i / R_i(0) = 1 + 2A_{2,i} \, C_i \, M_i \tag{3}$$

Here $(dn/dc)_i$ = specific refractive index increment of fraction i, and $K_{LS} = 4\pi^2 n_0^2 / \lambda_0^4 N_A$, where n_0 = refractive index of the solvent, λ_0 = wavelength of vertically polarized incident light in vacuum, N_A = Avogadro's number. The concentration-dependent terms in the right hand side of equations (2) and (3) with Huggins constant k_H and second virial coefficient A_2 (these two parameters can also depend on molar mass and vary across the chromatogram), reflect, respectively, hydrodynamic and thermodynamic polymer-polymer interactions in solution. If such interactions are negligible these two terms can be ignored and the corresponding values for intrinsic viscosity and molar mass are calculated as simple ratios of the detectors' responses:

$$[\eta]_i = (dn/dc)_i \, \eta_{sp,i} / \Delta N_i, \quad M_i = R_i(0) / K_{LS} (dn/dc)_i \, \Delta N_i \tag{4}$$

From equation (4) another important parameter, hydrodynamic volume H, can also be directly related to the detectors' responses (6):

$$H_i = [\eta]_i \, M_i = \eta_{sp,i} R_i(0) / K_{LS} \Delta N_i^2 \tag{5}$$

Hydrodynamic volume (5) can be expressed as a viscometric equivalent-sphere radius (or viscometric radius), $r_\eta = (3H/10\pi N_A)^{1/3}$, which reflects the hydrodynamic properties of a dilute polymer solution through its reduced shear viscosity. The hydrodynamic properties associated with the translational friction coefficient f (reciprocal of translational diffusion coefficient) can be described by another equivalent sphere radius, mean reciprocal radius (or hydrodynamic radius), $r_h = f/6\pi\eta_0$, $\eta_0 =$ solvent viscosity. This radius can also be measured on-line if a dynamic light scattering detector is a part of the multi-detector SEC arrangement (10). The geometrical radius (root-mean-square radius or radius of gyration) r_g is available from on-line static multi-angle laser light scattering (MALLS) experiments through the equation:

$$R_i(0)/ R_i(\theta) = P_i^{-1}(\theta; r_{g,i}) = 1 + 16\, r_{g,i}^2 \pi^2 n_0^2 \sin^2(\theta/2)\, /\, 3\lambda^2 \qquad (6)$$

where $R(\theta)$ and $P(\theta; r_{g,i})$ are the excess Rayleigh ratio and particle scattering factor at scattering angle θ and $\lambda = \lambda_0/n_0$. The function $P(\theta, r_{g,i})$, in a more general case than equation (6), depends on the conformational statistics of macromolecules (2). Finally, a thermodynamical-effective equivalent radius $r_T = (3A_2 M^2/16\pi N_A)^{1/3}$ can be constructed for consecutive elution fractions from corresponding Zimm-plots if several injections of different concentrations of the same polymer are made into a SEC system equipped with a MALLS detector (11).

All four molecular radii play important roles in elucidating the conformational properties of molecules with various architectures (2,12-16). Two types of combinations of these radii are usually used for estimation of polymer architecture: generalized (universal) ratios and contraction factors. The first are the ratios between different radii. Thus, the ratios $\phi = r_\eta/r_g$ and $\psi = r_T/r_g$ determine, respectively, the hydrodynamic draining function $\Phi = H/r_g^3 = 10\pi N_A \phi^3/3$ (also known as the Flory viscosity factor) and the thermodynamic segment-segment interpenetration function $\Psi = 4\psi^3/3\pi^{1/2}$ (2,14). Another universal ratio, $\rho = r_h/r_g$, reflects the sensitivity of the actual segment density to the hydrodynamic interaction and can be used, for example, for estimation of molecular asphericity of dendrimers (17). All three ratios, as well as other possible combinations of the molecular radii, are widely used in judging what architectural structure of macromolecules may be present (2,15,16).

Contraction (branching) factors represent the ratios of sizes of chemically similar branched and linear molecules with the same molar masses. Most popular are those for the radius of gyration and the viscometric radius: $g = r_{g,br}^2/r_{g,lin}^2$ and $g' = [\eta]_{br}/[\eta]_{lin} = r_{\eta,br}^3/r_{\eta,lin}^3$, which often serve as quantitative parameters in estimation of the degree of branching in stars, combs, and other branched molecules.

Theoretical Models

A number of theoretical equations for the aforementioned properties of macromolecules in dilute solution have been developed based on various model considerations (*1,2,12-17*) which always represented theoretical cornerstones of polymer science. Only recently, however, with the advent of contemporary synthetic methods capable of precise structural control of macromolecules with a variety of architectures (*18*), did polymer practitioners gain a real opportunity for experimental verifications of the theoretical predictions aimed at establishing important synthesis-structure-property relationships, as in the case of viscoelasticity and MMD studies of randomly branched polyesters (*19*). It is obvious that analytical methods such as multidetection SEC could and should play a decisive role in such development.

Multidetection SEC is very effective at calculating molar mass dependence for various sizes of macromolecules, e.g., through the conformation plot $r_g=K_gM^\nu$. Such power law behavior is typical of so-called self-similar objects, where an exponent ν describes the geometrical (fractal) dimension $d_f=1/\nu$ of the macromolecules (*2*). This dimension reflects the conformational behavior of macromolecules in dilute solution. Besides the topology of macromolecules it depends on their rigidity and interaction with the solvent. Thus, for linear chains d_f varies from 1 ($\nu = 1$) for rigid rods up to 3 ($\nu = 0.33$) for hard spheres, and has fractal values between 1.7 and 2 ($\nu = 0.588 - 0.5$) for coils depending on the thermodynamic quality of the solvent. The corresponding dependency, presented in a double logarithmic plot, results in a straight line with slope equal to the corresponding exponent. In principle, the fractal dimension of linear molecules can also be determined from the molar mass dependencies of other molecular radii, which at least asymptotically should have the same exponent ν as that of r_g. As a result, in the Mark-Houwink equation describing intrinsic viscosity as a function of molar mass the related exponent $\alpha_\eta = 3\nu-1$, in the power law describing the second virial coefficient A_2, the exponent $\alpha_T = 3\nu-2$, etc. What this means is that the ratios ϕ, ψ, etc., asymptotically tend toward some constants. These constants depend on solvent-polymer interaction, e.g. the degree of draining, and can be theoretically predicted from the various models (*1,2,12,14*). Hence, for linear macromolecules experimentally determined limiting values of these ratios as well as their molar-mass dependencies for intermediate M (*6,24*) help in elucidating the nature of thermodynamic and hydrodynamic interactions and their effect on chain conformations.

The situation is even more challenging for branched polymers. In general, d_f increases, while ν decreases, with increasing branching density. But now the fractal dimension can vary significantly for different parts of the same macromolecule. Thus, for polymers with long-chain branches the degree of shrinking (compared to the linear chain comprised of the same amount of monomer units) increases with the number of branching units contained in the selected portion of the branched molecule, even if the branching density λ (number of branching points per Dalton) is a constant across the entire molecule. Quantitatively, this is described by the following equations for the branching contraction factor g, developed for randomly branched Gaussian chains with functionality of branching units of 3 and 4, respectively (20):

$$g_3 = [(1 + \lambda M /7)^{1/2} + 4\lambda M/9\pi]^{-1/2}, \quad g_4 = [(1 + \lambda M /6)^{1/2} + 4\lambda M /3\pi]^{-1/2} \quad (7)$$

Even in a simple star polymer with a few arms the region close to a branching point is denser and has higher fractal dimension than the outer, more dilute region. With an increase in number of arms f, the scaling theory predicts the existence of three characteristic regions, the inner melt-like extended core region, the intermediate region resembling a globule with uniform density, and the outer semi-dilute region where each arm constitutes a series of blobs with sizes increasing in the direction of the arm end (15). Each region has a specific fractal dimension, which is reflected in the corresponding exponent v in the molar mass dependencies of the molecular radii measured by SEC.

Star-branched macromolecules with arms long enough to belong to the most dilute region have the same exponent as their individual arms, i.e. linear molecules, and the corresponding double-logarithmic plot is parallel to that of linear chains, but shifted to the lower values as the number of arms increases (21). Such shift for the conformation and Mark-Houwink plots is described by the corresponding contraction factors, g and g', which in this case are determined exclusively by the number of arms. Several calculations for different types of stars proposed (2, 12-15, 20) for g (much fewer and more empirical by nature for g') can be successfully tested by multidetector SEC if molar mass dependencies of the molecular radii of corresponding linear molecules can be independently measured. A similar approach can also be used for comb-like polymers (22) or for polymers with short-chain branches, e.g., copolymers of ethylene with α-olefins, where both g and g' are also molar-mass independent.

Contrary to linear molecules, the different radii of branched molecules are affected by branching in different ways, so that they may have different exponents in the power law. Thus, the ratios ϕ and ψ depend, respectively, on the solvent-polymer and polymer-polymer repulsive interactions, which have a stronger effect in branched molecules, so that both ratios increase with branching density. There is no conclusive theory describing the relationships between geometrical and hydrodynamic radii of branched molecules, e.g., between g' and g, and multidetection SEC thus becomes a unique source of valuable information needed for validation of different semi-empirical approaches.

Improvements in Data Reduction Methodology

Hydrodynamic Volume and the Universal Calibration Concept

The ability to calculate different radii helps in answering the fundamental question in SEC: which size, if any, controls the separation? A commonly accepted universal calibration concept (*3*) points at the hydrodynamic volume H. In other words, the elution volume dependence of log H (the so-called hydrodynamic volume calibration curve) is considered as "universal" (polymer-independent), and triple detection allows for a direct verification of this concept by comparing slice values H_i from equation (5) with the corresponding "standard" (peak) values $H_{st,i}$ obtained from a set of narrow polydispersity standards with different molar masses. Although with triple detection the question loses at least part of its topicality because of eliminating the need for external universal calibration to calculate the properties listed in the previous section, such comparison may play an important role in evaluating the possible non-size-exclusion effects, some of which can significantly reduce the accuracy of the calculations, as well as in establishing the accurate values of system and sample-dependent parameters (*6*).

Non-Ideality of Size-Exclusion Separation

There are four major reasons for the "non-ideal" behavior of polymers subjected to size-exclusion separation, provided all system parameters such as interdetector volume(s), instrument calibration constants, etc., are accurate: polymer-polymer interaction, non-steric interaction between a polymer and a stationary phase, band-broadening, and compositional and/or architectural heterogeneity of macromolecules. Each of these factors results in deviation of apparent H-values, calculated from equation (5), from the H_{st}-values corresponding to the ideal separation. Additionally, some publications report a failure of universal calibration for specific types of conformations, such as rod-like molecules (*23*). In our SEC lab we use equation (5) as a routine test of all analyzed polymers, which include many types of conformations, e.g., rod-like liquid-crystal polyesters, and did not observe deviation caused by any reason different from the four above-listed factors. Nevertheless, as the theoretical justification of universal calibration is still a matter of discussion, one can assume the possibility of such a violation.

The effect of non-ideality of separation on deviation from universal calibration can be characterized by a non-ideality factor $P_i = (H_i/H_{st,i})$. The detailed analysis of all four cases was published recently (6), where accurate equations for P as a function of elution volume V were developed for each case. Here we present a brief review with emphasis on the effect of different types of non-ideality on the accuracy of polymer characterization by multidetector SEC.

Polymer-Polymer Interaction. If the concentration of a polymer solution is not low enough for the concentration terms in equations (2) and (3) to be ignored, a weak intermolecular interaction may impact both separation and detection, and the non-ideality factor in the first approximation can be presented as a difference between thermodynamic and hydrodynamic contributions:

$$P_i = 1 + 2\ [A_{2,i}R_i(0)/\ K_{LS}(dn/dc)_i^2 - k_{H,i}\eta_{sp,i}] \tag{8}$$

Since both coefficients A_2 and k_H are positive for typical SEC conditions, the thermodynamic interaction (interchain repulsion) increases the effective size of macromolecules but the hydrodynamic interaction (draining effect) decreases it. Accurate calculation of the polymer properties using equations (1)–(3) is still possible but represents a challenging problem, as both parameters should be known across the entire polymer molar mass distribution, and the second virial coefficient can noticeably depend on molar mass, especially for branched molecules (2). It was suggested recently (11) that several injections of different concentrations of the same sample into a SEC system with MALLS detection allows for calculation of A_2 using a Zimm plot. A similar approach with a viscometer detector may be used to calculate k_H.

In the case of thermodynamically poor solvent, the direct consequence of intermolecular interaction can be aggregation, which distorts the actual MMD of the polymer and its dilute solution properties. Small aggregates, which can make their way to the detector cell, are easily identified by light scattering, but resolution can be lost if the size of the aggregates exceeds the exclusion limit of the columns. Additional complication can be caused by the strong, sometimes irreversible retention of aggregates due to adsorption onto the separation media. The case of polyelectrolytes, where both inter- and intramolecular interactions may significantly complicate the interpretation of light-scattering measurements, is not considered in this paper.

Non-Steric Interaction. A consequence of non-steric interaction between solute and stationary phase could be a shift of the elution volume to higher values as compared to separation by a size-exclusion mechanism only (attractive interaction) or, less likely, to lower values in the case of repulsive forces (e.g., electrostatic interaction in aqueous eluents with high dielectric constant).

Respectively, positive (P > 1) or negative (P < 1) deviations from standard universal calibration are observed, which are much stronger for the higher molar mass fractions. In other words, such interactions lead to positive (clockwise) or negative (counter-clockwise) rotations of the hydrodynamic volume (5) around the point corresponding to the total liquid volume of the columns.

In most cases, equations (4) and (6) are still valid in calculating polymer solution properties in the presence of a weak non-steric interaction, but the separation power of the columns, as well as their efficiency can be significantly reduced. High molar-mass particles, e.g., highly branched polymers, can elute together with much lower molar mass linear molecules creating a local polydispersity, or irreversibly adsorb onto the columns reducing sample recovery. In any case, a quantitative estimation of adsorption or other non-steric interaction effects can be done by measuring the non-ideality factor P across the entire elution profile.

Band Broadening. Due to the limited efficiency of the chromatographic separation, each elution volume V_i still contains macromolecules with different sizes and, hence, with different molar masses and intrinsic viscosities. As a result of such local heterogeneity, instead of instantaneuous values on-line detectors produce weight-averages of the intrinsic viscosity and molar mass in equations (4), and z-average geometrical and hydrodynamic radii, for the local distributions within each slice. This effect can significantly compromise the accuracy of calculations, especially those related to the structure of macromolecules such as the conformation plot, even for samples with broad MMD.

Some mathematical correction of this problem is possible by extracting the instantaneuous values of the polymer properties from the averaged values for each elution volume, i.e., by constructing "true" (instantaneuous) calibration curves from those obtained from direct on-line measurements. This axial dispersion correction could be done for example by iterative deconvolution (*3*) applied to measured multi-detector chromatograms if the corresponding spreading function responsible for band broadening is known. It is important to note that usually this function is much broader at lower elution volumes (higher molar masses) because of the nature of the band-broadening phenomenon in polymer chromatography: diffusion of macromolecules inside pores, which strongly depends on molar mass. An additional contribution comes from a general smoothing procedure in SEC: equal increments of elution volume in the high molar mass region contain more fractions than those located downstream due to the near-exponential dependence of molar mass and radii on elution volume.

If the standard (not affected by band broadening) universal calibration, H_{st} versus V, is established the calculation of the apparent hydrodynamic volume H from (5) allows for accurate estimation of the spreading functions at each elution volume provided there are no other potential sources of non-ideality (6). Band broadening rotates H in a counter-clockwise direction about the point corresponding to the apex V_{apex} of the concentration chromatogram, that is P < 1 if V < V_{apex} and P > 1 if V > V_{apex}, and the P-values determine the local polydispersity at each elution volume.

Polymers with Compositional or Topological Heterogeneity. Another possible contributor to local polydispersity is the structural or compositional heterogeneity of a polymer, as macromolecules with different molar masses may have the same size due to differences in chemical composition or molecular architecture (branching). As a result, each elution volume can be associated with some local molar mass and intrinsic viscosity distributions. Weight-average moments of these distributions are directly calculated from equations (4). In this case, the apparent hydrodynamic volume H from equation (5) always exceeds the actual value H_{st}, and their ratio P gives the polydispersity of the local polymer distribution in each slice i, so that

$$P_i = M_{w,i} / M_{n,i} = [\eta]_{w,i} / [\eta]_{n,i} \qquad (9)$$

where subscripts 'w' and 'n' stand for weight- and number-average, respectively.

In the case of copolymers or polymer blends, individual macromolecules can differ in refractive index increment. Equation (9) is still valid if such a difference exists only between different slices of the elution profile, while inside each slice all macromolecules are "isorefractive." If this is not true, the measured non-ideality index can noticeably exceed the local polydispersity when the difference between the refractive index increments of macromolecules is significant (for example, the same elution volume has macromolecules with positive and negative dn/dc). In this case light scattering does not provide the true average molar mass in each slice (25), and more accurate results could be obtained using dual detection SEC (refractometer and viscometer) with universal calibration. A more comprehensive solution for the polymer characterization in this case is a two-dimentional approach, where a polymer sample is separated first by composition using one of several interaction chromatography modes, and then compositionally homogeneous fractions are subjected to multi-detector size-exclusion separation (26).

System (Sample) Parameters

An accurate determination of multi-detector system (instrument calibration constants, interdetector volumes, etc.) and sample (concentration, refractive index increment, etc.) parameters is an important part of a successful application of the SEC method to polymer characterization. Such determination is not always straitforward (27). For example, an apparent interdetector volume measured as a difference between apexes of corresponding chromatographic peaks for a monodisperse sample can be affected by instrument band broadening. Internal consistency tests can greatly help in overcoming such problems. Thus, many whole (bulk) polymer properties such as the weight-average molar mass or intrinsic viscosity can be measured independently by integration of single detector traces. A systematic approach has been developed (27) to diagnose and troubleshoot these parameters using comparison between different calibration curves obtained for well-characterized polymer standards. The hydrodynamic volume calibration from equation (5) can serve as an additonal tool in this approach.

Let us consider one example. If the light sources for the refractive index and light scattering photometers have different wavelengths, the corresponding increments $(dn/dc)_{DR}$ and $(dn/dc)_{LS}$ in equations (1) and (3) respectively may differ also, because dn/dc depends on wavelength. If universal calibration values, $H_{st,i}$, are known, then

$$(dn/dc)_{LS,i} /(dn/dc)_{DR,i} = (\eta_{sp,i} R_i(0)/ K_{LS} \Delta N_i^2 H_{st,i})^{1/2} \quad (10)$$

For many polymer-solvent combinations, it is not easy to find published dn/dc values for different wavelengths. Equation (10) allows for calculating the refractive index increment for one photometer if the corresponding value for another photometer with different wavelength is known and the universal calibration curve is established. This will provide increased accuracy in the estimation of polymer properties from multi-detector SEC.

Elimination of Concentration Detector

The use of a differential refractometer or any other concentration detector sometimes limits the applicability of multi-detector SEC for characterization of polymers with a small portion of high-molar-mass material. Two new approaches were proposed recently (6,28) which allow calculations to be performed without using the concentration detector trace. Thus, the light scattering detector signal can replace the refractometer detector signal using the relationship $\Delta N_i \sim R_i(0)/M_i$ to calculate higher-order molar mass averages of polymers (28). This approach still requires accurate molar mass values M_i for each slice, i.e., a molar mass calibration curve. The use of external standards proposed by these authors (28) avoids the need to apply equations (4) with the refractometer's traces, but will limit the applicability of calculations to linear polymers with a structure similar to that of standards. In the next section we describe a novel computational algorithm to construct such curves without external standards, which can augment the method (28) to include branched and other polymers with complex structures.

Another way to avoid a concentration detector is to replace ΔN in equations (1)–(5) with the hydrodynamic volume H (6). In this way MMD, intrinsic viscosity, and other polymer solution properties can be determined without a concentration detector if the universal calibration curve is known either from external standards or from the procedure described in the next section. This method can be very useful for polymers with long-chain branching or polymer blends, especially in calculating structural parameters affecting rheological properties of polymer melts (8,9,17,28).

Model Fittings in Multi-Detector SEC

Additive detector noise, seen as random fluctuations in the baseline having zero mean and well-defined standard deviation, is an irreducible component of the measurement process. As can been seen from equations (4) and (5) the slice values of molar mass, intrinsic viscosity and hydrodynamic volume are proportional to the ratios of the detector responses. It follows that detector noise introduces the non-random noise in quantities that depend on these ratios. At the tails of distributions, these ratios do not produce physically meaningful values due to the differences in sensitivities between molar mass and concentration detectors. Fitting a smooth, multi-variate model to a time series of noisy data is generally an effective way to produce a more precise estimate of the measured quantity at each sample time (29). In the case of triple detection SEC the most frequently used models are the molar mass and intrinsic viscosity calibration curves, such as the low-order (usually, between 1 and 5) polynomial F_M in equation

$$\log M_{fit} = F_M (V) \equiv C_{M,0} + C_{M,1}V + \ldots + C_{M,s}V^s \qquad (11)$$

where the fitting parameters $C_{M,0},..,C_{M,s}$ describing the calibration curve depend on both the polymer structure and the separation system (columns, mobile phase, temperature).

Traditionally, a linear least-squares procedure is used to fit a polynomial to the data by minimizing the following expression

$$\chi^2 = \Sigma(\log M_i - F_{M,i})^2 \qquad (12)$$

In this expression, the values for log M_i are the slice measurements from equations (4), $F_{M,i} = F_M(V_i)$ is the model slice values for log M_{fit}, and the sum is over the individual slices of the polymer peak. The least-squares fit (12) cannot be used at the tails of the distribution because the logarithm of a ratio of measurements, one of which fluctuates near zero, creates a non-random infinitely large biased noise. For this reason the least-squares fitting of the model is confined to the "heart" of the peak where the signal-to-noise ratios for the measured quantities are high. This truncation is followed by the extrapolation of the model results to the polymer tails. Though critical to obtaining the MMD and polymer solution properties for the entire polymer distribution, the extrapolation procedure is problematic. The results are extremely sensitive to the choice of the demarcation between the heart and the tails of the peak and to the method of its implementation (28), especially for polymers with tailed distributions.

A more accurate approach to least-squares fitting (30,31) uses the same model (11) for the calibration curve, but directly compares the measurements of R(0) to a *model* of R(0). The quantity to be minimized is then

$$\chi^2 = \Sigma[R_i(0) - \Delta N_i \, K_{LS}(dn/dc)10^{F_{m,i}}]^2 \qquad (13)$$

Equation (13) is a rearrangement of the same quantities used in equation (12), but with important advantages. Because the errors in R(0) and ΔN are assumed to have zero mean and distributed as Gaussian, the error in each residual term, $R_i(0) - \Delta N_i \, K_{LS}(dn/dc)10^{F_{m,i}}$, will also have zero mean and be distributed as Gaussian. All the chromatographic profile data in the fit can be used, including points for which the measured values for $R_i(0)$ and ΔN_i are consistent with zero and in fact have negative values due to noise. The above expression (13) can be used to fit data throughout the entire polymer region, including the tails, which may provide useful information to constrain the least-squares fit. The proper choice of a weighting factor (the variance of each term in equation (13)) can further improve the least-squares fitting (30). Least-squares procedures similar to (13) can be applied to construct the smooth intrinsic viscosity and hydrodynamic volume (universal) calibration curves based on definitions (4) and (5).

The radius of gyration calibration curve,

$$\log r_{g,fit} = F_g(V) = C_{r,0} + C_{r,1}V + \ldots + C_{r,s}V^s \qquad (14)$$

can be constructed if MALLS is used as an on-line detector. Traditionally, these calculations are performed in two steps: first, each slice value $r_{g,i}$ is independently determined as a slope when extrapolating expression (6) to zero angle, and then these slice values are fitted by polynomial (14). This approach suffers from the same drawback as that of equation (12): slices with low signal-to-noise ratio adversely affect the calibration curve. Much more accurate results can be obtained (32) if the calibration curve is calculated in one step: by minimizing expression

$$\chi^2 = \sum[R_i(\theta_j) - R_i(\theta^*)P(\theta_j; 10^{F_{g,i}}) / P(\theta^*; 10^{F_{g,i}})]^2 \qquad (15)$$

The asterisk here refers to one of the measured angles, and sum is over *all* slices i and measured angles θ_j, so that the calibration curve (14) as well as the fitted particle scattering function $P(\theta; r_{g,fit})$ are determined through the measured quantities only and without data extrapolation or truncation. The molar mass calibration curve (11) can then be obtained by minimizing expression

$$\chi^2 = \sum[R_i(\theta_j) - \Delta N_i\, K_{LS}(dn/dc)\, P(\theta_j; 10^{F_{g,i}})\, 10^{F_{m,i}}]^2 \qquad (16)$$

which produces more accurate results than equation (13) in the case of MALLS without zero angle.

This approach, which we call a calibration fit, constitutes an intermediate step in determining the MMD and polymer solution properties. Such properties can then be calculated by excluding elution volume from the corresponding calibration curves, which requires an additional fit between related properties, e.g., between radius of gyration and molar mass (conformation plot). This last step (a structural fit) is done using either an empirical polynomial or a proposed physical model developed for polymers with specific configurational and conformational structure, such as described by equations (7). The accuracy of the structural fit will greatly depend on the accuracy of the related calibration curves and may be improved if performed in a single step. For example, the exponent v (or fractal dimension $d_f = 1/v$) in power law $r_g = K_g M^v$ for linear polymers can be determined directly by minimizing expression

$$\chi^2 = \sum[R_i(\theta_j) - \Delta N_i\, K_{LS}(dn/dc)\, P(\theta_j; 10^{F_{g,i}})\, (K_g\, 10^{\,F_{g,i}})^{1/v}]^2 \qquad (17)$$

regarding a single parameter v, thus avoiding the construction of the molar mass calibration curve (11) with expressions (13) or (16). Such approach can also be applied for polymers with more complex architecture, e.g., branched polymers described by model (7), but prior knowledge about polymer structure is essential to avoid inappropriate model assumptions.

Characterization of Ethylene/Methyl Acrylate Copolymer

To illustrate the preceding the commercial ethylene/methyl acrylate statistical copolymer (EMA) Optema™ TC113 (25% methyl acrylate) from Exxon (Houston, TX) was characterized by a triple detector SEC Alliance™ GPCV 2000 system from Waters Corporation (Milford, MA) equipped with a differential refractometer as a concentration detector, a differential viscometer, and two-angle (15° and 90°) light scattering photometer PD2040 from Precision Detectors (Franklin, MA). All measurements were performed at 150 °C in 1,2,4-trichlorobenzene stabilized with 0.025% 2,6-di-*tert*-butyl-4-methyl phenol. The Waters Empower™ triple detection software with customized options to incorporate the above approaches was used for data reduction.

The overlay of elution traces from all three detectors (Figure 1) indicates a significant polydispersity of the sample with a noticeably lower retention time shoulder typical of the products of a high-pressure free-radical autoclave process. Long-chain branches as well as possible chemical heterogeneity may complement to the local polydispersity of the separated fractions. Calculated across the polymer distribution (after the appropriate axial dispersion correction), the non-ideality factor P shows a negligible effect on the measured hydrodynamic volume compared to that obtained from a set of narrow polydispersity polystyrene standards (Figure 2), which indicates a homogeneous distribution of comonomer in the polymer chains. A slight increase in hydrodynamic volume at the highest molar mass region is probably due to significant branching in these fractions.

Using least-squares fitting criterion (15), (16) for r_g and M respectively, and similar expressions for $[\eta]$ and r_η, four calibration curves have been constructed (Figure 3). Non-random noise at the tails of the polymer distribution (especially that of log r_g at higher retention times) does not affect the fitted curves due to the selected computational algorithm. Notice that radius r_η increases with molar mass much faster than does r_g, so that their ratio $\phi = r_\eta/r_g$ exceeds unity at higher molar masses. This behavior is known for branched polymers (24) and is opposite to that observed recently for linear molecules (6).

Figure 4 demonstrates the calculated MMD and structural curves reflecting polymer solution properties, such as conformation and Mark-Houwink plots. The initial slopes of these curves, $\nu = 0.56$ and $\alpha_\eta = 0.69$, describing a hypothetical linear polymer with the same backbone, are calculated directly from the fitted data without any prior information about the polymer. Slightly lower values of these exponents, as compared to those for polyethylene (30), can be explained as the effect of polar comonomer in non-polar solvent. The results unambiguously demonstrate a significant amount of long-chain branches in the macromolecules as can be seen from the branching contractions factors g and g′ calculated across the polymer distribution (Figure 4).

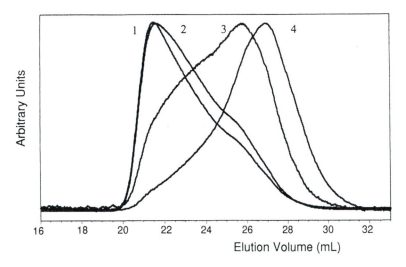

Figure 1. Normalized chromatograms from three on-line detectors: light scattering at 15° (1) and 90° (2), viscometer (3) and refractometer (4).

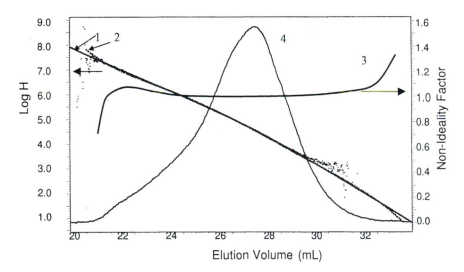

Figure 2. Comparison between apparent (1) and "standard" (2) hydrodynamic volume calibration curves obtained with 3rd order polynomial fit to slice values (dots) from equation (5) and peak values of narrow polystyrene standards, respectively. 3 – non-ideality factor P, 4 – concentration chromatogram.

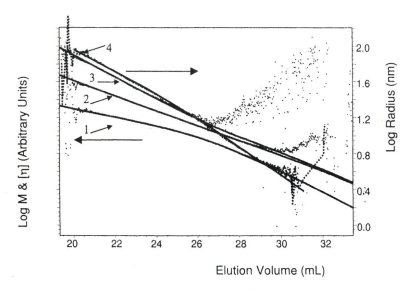

Figure 3. Calibration curves obtained with 3rd order polynomial for intrinsic viscosity (1), molar mass (2), radius of gyration (3) and viscometric radius (4). Dotted lines – direct slice measurements.

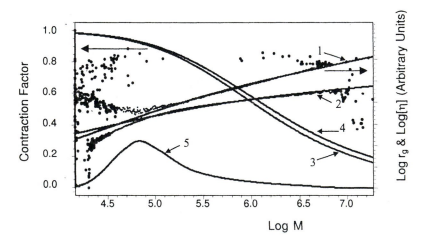

Figure 4. Conformation (1) and Mark-Houwink (2) plots, branching contraction factors g (3) and g' (4), MMD (5)

Different theoretical models can be tested based on these fitted curves. Thus, following the assumption of randomly branched Gaussian chains (20), the branching density λ from equation (7) is calculated for each elution volume V. The accuracy of such calculations depends on the accuracy of the calibration curves. If we assume that λ is a constant across the entire polymer distribution (the basic assumption in the Zimm-Stockmayer model (20)), then the calibration curves can be completely avoided and an expression similar to equation (17) for linear polymers is used to determine just three parameters: $K_g = 0.0258$ nm, $\nu = 0.56$ and $\lambda = 3.5 \times 10^{-5}$. Similar approach can be used to determine a structural model characteristics of the hydrodynamic properties of the polymer solution, such as the exponent ε in the relationship $g' = g^{\varepsilon}$. The structural least-squares fit with the assumption that this exponent is the same for all branched fractions gives $\varepsilon = 0.85$, which is a typical value for randomly branched polyethylenes.

Conclusions

New data reduction algorithms significantly expand the polymer characterization capabilities of SEC coupled with molar mass-sensitive detectors. Quantitative characterization of non-size-exclusion factors with a hydrodynamic volume calibration curve not only allows for a better understanding of the mechanism of separation but also provides a practical tool for the estimation of the effect of non-ideality of separation on the accuracy of the calculations. Elimination of concentration detector traces significantly improves quantitation of polymers with long high-molar-mass tails. Proper smoothing and fitting procedures in calculating the MMD and structural characteristics of macromolecules in dilute solution may be a key factor in elucidating polymer solution properties by multi-detector SEC. The computational algorithms presented for molar-mass-sensitive detectors could also be used for other multi-detector combinations.

Literature Cited

1. Flory, P. J. *Principles of Polymer Chemistry*, Cornell University Press: Ithaca, NY, 1953.
2. Burchard, W. Solution Properties of Branched Macromolecules. *Advances in Polymer Science*, Roovers J., Ed.; Springer: Berlin, 1999; Vol.143, pp 113-194.
3. Mori, S.; Barth, H. *Size-Exclusion Chromatography*, Springer: Berlin, 1999.

300

4. Colby, R. H.; Rubinstein, M.; Gillmor, J. R.; Mourey, T. H. *Macromolecules* **1992**, *25*, 7180.
5. Weissmuller, M.; Burchard, W. *Polymer Internat.* **1997**, *44*, 380.
6. Brun, Y. *J. Liq. Chromatogr.* **1998**, *21*, 1979.
7. Balke, S. T.; Mourey, T. H. *J. Appl. Polym. Sci.* **2001**, *81*, 370.
8. Monfort, J. P.; Marin, G.; Monge, P. *Macromolecules* **1988**, *19*, 1979.
9. Eder, G.; Janeschitz-Kriegl, H.; Liedauer, S.; Schausberger, A.; Stadlbauer, W.; Schindlauer, G. *J. Rheol.* **1989**, *33*, 805.
10. Cotts, P., this book.
11. Girod, S.; Baldet-Dupy, P.; Maillots, H.; Devoisselle, J.-M. *J. Chromatogr. A* **2001**, *943*, 147.
12. Freire, J. J. *Advances in Polymer Science*; Roovers J., Ed.; Springer: Berlin, 1999; Vol. 143, pp 35-112.
13. Roovers, J. *Dendrimers and other Dendritic Polymers*; Frechet, M. J; Tomalia, D. A., Eds.; John Wiley & Sons: Chichester, UK, 2001; pp 67-90.
14. Freed, K. F. *Renormalization Group Theory of Macromolecules*; John Wiley & Sons: New York, 1987; ch.10.
15. Grest, G. S.; Fetters, L. J.; Huang, J. S. *Advances in Chemical Physics;* Prigogine, I.; Rice, S. A., Eds.; John Wiley & Sons: New York, 1996; Vol. XCIV, pp 67-163.
16. Striegel, A., M; Plattner, R., D.; Willett, J. L. *Anal. Chem* **1999**, *71*, 978.
17. Dvornic, P. R.; Uppuluri, S. *Dendrimers and other Dendritic Polymers*; Frechet, M. J; Tomalia, D. A., Eds.; John Wiley & Sons: Chichester, UK, 2001, pp 331-358.
18. Robello, D. R.; Andre, A.; McCovick, T. A.; Kraus, A.; Mourey, T. H. *Macromolecules* **2002**, *35*, 9334.
19. Lusignan, C. P.; Mourey, T. H.; Wilson, J. C.; Colby, R. H. *Phys. Rev. E* **1999**, *60*, 5657.
20. Zimm, B. H; Stockmayer, W. H. *J. Chem. Phys.* **1949**, *17*, 1301.
21. Roovers, J.; Zhou L.-L.; Toporowski, P. M.; van der Zwan, M.; Iatrou, H.; Hadjichristidis, N. *Macromolecules* **1993**, *26*, 4324.
22. Radke, W.; Muller, A. H. *Polymer Prepr.* **1999**, *40*, 144.
23. Dubin, P. L.; Principi, J. M. *Macromolecules* **1989**, *22*, 1891.
24. Wintermantel, M.; Antonietti, M.; Schmidt, M. *J. Appl. Polym. Sci.: Appl. Polym. Symp.* **1993**, *52*, 91.
25. Kratochvil, P. *Classical Light Scattering from Polymer Solutions*; Elsevier: Amsterdam, 1987; ch. 5.
26. Pasch, H; Trathnigg, B. *HPLC of Polymers*; Springer: Berlin, 1998.
27. Mourey, T.H.; Balke S.T. A Strategy in Interpreting Multidetector Size-Exclusion Chromatography Data I. *Chromatography of Polymers: Characterization by SEC and FFF*, Provder T., Ed.; ACS Symposium Series: Washington, DC, 1993; pp 180-198.

28. Yau, W. W.; Gillespie, D. *Polymer* **2001**, *42*, 8947.
29. Lew, R.; Cheung, P.; Balke, S. T; Mourey, T. H. *J. Appl. Polym. Sci.* **1993**, *47*, 1700.
30. Brun, Y.; Gorenstein, M. V.; Hay, N. *J. Liq. Chromatogr.* **2000**, *23*, 2615.
31. Gorenstein, M. V; Brun, Y. *USA Patent 6064945*, **2000**.
32. Rong, X; Brun, Y; Gorenstein, M. V. *Patent Pending*, **2001**.

Chapter 18

Using Multiple Detectors to Study Band Broadening in Size-Exclusion Chromatography

Miloš Netopilík

Institute of Macromolecular Chemistry, Academy of Sciences of the Czech Republic, 162 06 Prague 6, Czech Republic

This chapter reviews the sources of band broadening, including the separation process itself, restricted to the case of a chromatographically-simple polymer and Gaussian band-broadening function. Methods of determining band broadening using multiple detection are also reviewed, as based on the latest contributions of the author to the problem.

Size-exclusion chromatography (SEC) is a method of separation of polymers according to molecular weight where, in the case of chromatography-simple polymers the latter is related to their hydrodynamic volume (*1*). Soon after invention of the method (*2*) mass detection of the polymer, frequently with a refractive index (RI) detector, was supplemented by viscometric (*3-5*) and light-scattering (*6-8*) detection.

Chromatographic separation of polymers is possible in the adsorption mode, where the molecules of a polymer (analyte) in the mobile phase (MP) are temporarily adsorbed on (or react with) the solid phase (SP) by van der Waals forces. In SEC mode, due to their thermal motion the molecules of the analyte (analyzed polymer) are temporarily captured in the pores of the SP with the capture probability determined by the size of the analyte and by the accessible portion of the inner pore volume, or in the combination of the two (adsorption and size-exclusion) modes (*9*). In the SEC mode, the flux of the analyte between the phases is caused by the difference, produced by flow of the MP, in the entropic part of the chemical potentials between the phases (*10*). In both modes, the fractionation is accomplished by the transport of the molecules in the MP and by the delay of molecules in/on the SP. The mechanism of chromatographic separation was described using a kinetic formalism based on the probabilities of adsorption and desorption (*11*). This description has been developed and is suitable for the adsorption mode. The SEC mode was described by postulating an equilibrium of the analyte in the MP and in the pores of the SP in a close region (*12*), combined with displacement. Both descriptions were proven to be equivalent (*13*). From the theoretical analysis it follows that the separation process is one of the sources of band broadening (*11,13*), a process which decreases the resolution power of the separation system (*14*). However, in the case of an effective separation system and an analyte with medium-broad molecular weight distribution (MWD), the experimental error caused by band broadening is comparable with errors caused by other sources of error (*15*) and in practice is frequently neglected.

This review, based mainly on the author's contributions, presents the separation and detection processes from the viewpoint band-broadening (axial dispersion) of chromatographically-simple analytes. The interactions of a molecule of such analyte with the SP can be described (*12*) by the mean fraction of analyte in the MP in a restricted region (plate) forming a part of the separation system (column) from the total amount in the MP and the SP. Equivalently (*13*), the interactions can also be described by the probabilities per unit time of adsorption and desorption on/from the SP. Both approaches lead to a band-broadening function which is principally skewed and tends, with increasing number of interactions of the analyte molecules with the SP, to the symmetrical Gaussian distribution and that describes basic features of band broadening. This agrees qualitatively both with experiment (*16*) and with the approach based on the distribution of times spent by molecules in pores of the SP (*16*).

If only one detector is employed, band broadening can be estimated only indirectly. An elution curve identical to the spreading function, the broadness of which is a measure of band broadening, is obtained only for an analyte of uniform in molecular weight, M, which is usually a low-molecular-weight (M) substance (*e.g.,* toluene) in organic-phase separation systems or proteins (*17*) in aqueous MP, and its characteristics can be determined directly. In organic MP, however, only reference standards with MWD, but not uniform in M, are available. To characterize their MWD the extent of band broadening must be known precisely. Absolute methods, such as sedimentation equilibria, used for the determination of the weight-to-number-average molecular weight ratio, $\overline{M}_w / \overline{M}_n$, which is used for checking the SEC results (*2*) are now obsolete and there is a lack of methods for $\overline{M}_n$ and $\overline{M}_w / \overline{M}_n$ determination. Therefore, the potential of multidetector SEC is to be used to get all information from the SEC data. This is possible preferably using multiple detection (*18-22*).

The analysis of a system with a dual detection can yield, in principle, two values of the $\overline{M_w}/\overline{M_n}$ ratio. Their difference is a measure of the band broadening of a particular system. The band broadening obtained using the calibration dependence, i.e., the logarithmic dependence of M on elution volume (V), in a wide range of V (and M) is overestimated. On the contrary, the band broadening obtained from a local calibration dependence, i.e., obtained from the dual detection of M by the combination a light scattering (LS) with a concentration-sensitive detector (differential refractometer (8)) is underestimated. The correct $\overline{M_w}/\overline{M_n}$ ratio lies between the two values and can be assessed by a correction procedure. The correction method can be based (23) either on any numerical (point by point) deconvolution procedure of the elution curves (EC) (14,24-26) or on the correction of the molecular weight averages calculated from the concentration EC and a broad-range calibration dependence (30) and from the dual detector record (22). As the latter method is based on the approximation of the sample MWD by the log-normal function, it is used preferably on analyses of narrow-MWD samples where the point-by-point methods tend to instability (25) and the correction of the dual detector record is prone to fail (27). This makes SEC with dual or multiple detection an excellent tool for studying band broadening and characterization of the MWD of such samples (22). In the following sections band broadening will be discussed from the viewpoint of multiple detection.

Formulation of Band Broadening

Band broadening is described by an equation proposed by Giddings and Eyring (11) and frequently referred to as the Tung (14) equation

$$F(V) = \int_{-\infty}^{\infty} W(y)G(V,y)\,\mathrm{d}y \tag{1}$$

relating experimental and theoretical elution curves, $F(V)$ and $W(y)$, respectively, where $G(V,y)$ is the band-broadening (spreading) function, which is the (hypothetical) elution curve of the individual species uniform in M (it is identical with the elution curve only for analytes uniform in M). This function of two variables for elution volume, V and y, gives the contribution to the experimental elution curve $F(V)$ at elution volume V of a fraction (band) of polymer $W(y)\mathrm{d}y$ with molecular weight M related to elution volume by an equation called 'calibration dependence', in the first approximation linear,

$$\ln M = A + By \tag{2}$$

where A and B are constants and y is the elution volume of the maximum of the spreading function (12)

$$y = V_0/p \tag{3}$$

where V_0 is the excluded volume of the separation system and p the mean fraction of the analyte in the MP from the total amount in the MP and in/on the SP.

Solving equation (1) means finding $W(y)$ from the known $F(V)$. It is called the inverse problem and several mathematical methods were developed for this purpose (14,24-26). Equation (1) can be solved analytically under the condition that the spreading function is approximated by the (Gaussian) normal distribution (28,29)

$$G(V, y) = \frac{1}{\sigma\sqrt{2\pi}} \exp\left[-\frac{(V-y)^2}{2\sigma^2}\right] \tag{4}$$

with variance σ^2 and the MWD is approximated by log-normal function

$$w(M) = \frac{1}{\beta M \sqrt{\pi}} \exp\left[-\frac{1}{\beta^2} \ln^2 \frac{M}{M_0}\right] \tag{5}$$

where $M_0 = \sqrt{M_w M_n}$ and $\beta = \sqrt{2 \ln M_w / M_n}$ are parameters characterizing the position and broadness of the distribution. (The variance of the log-normal MWD is given by $\sigma_{MWD}^2 = \beta^2/2$.) The experimental elution curve is then (14)

$$F(V) = -\frac{B}{\beta} \sqrt{\frac{\Sigma}{\pi}} \exp\left[-\frac{B^2 \Sigma}{\beta^2} (y - V_0')^2\right] \tag{6}$$

where $V_0' = (\ln M_0 - A)/B$ (cf. equation (2)) and

$$\Sigma = \frac{\beta^2/B^2}{2\sigma^2 + \beta^2/B^2} \tag{7}$$

For the weight-to-number-average molecular weight ratio (calculated using M given by calibration dependence 2, subscript 'c'), Hamielec derived, without any assumption concerning the MWD (30), the formula

$$\left(\overline{M}_w / \overline{M}_n\right)_c = \left(\overline{M}_w / \overline{M}_n\right) \exp[\sigma^2 B^2] \qquad (8)$$

The spreading function as a result of interaction of the analyte with the SP was first expressed in terms of the elution time, for an analyte characterized by probabilities of adsorption, k, in the MP, and desorption in/on the SP, k', as (*11*)

$$P(t) = \frac{(kk't^o)^{1/4}}{2\sqrt{\pi}} \frac{1}{t^{3/4}} \exp\left[-\left(\sqrt{k't} - \sqrt{kt^o}\right)^2\right] \qquad (9)$$

where t^o is the elution time corresponding to the exclusion volume, V_0.

Equation (9) was derived by asking by what time, t, a molecule is delayed on the SP, i.e., at what time does the molecule arrive at the end of the column after the time t^o? The elution time from the injection

$$t_c = t^o + t \qquad (10)$$

is related to the elution volume by

$$t_c = V/r_f \qquad (11)$$

where r_f is the flow rate.

The theory of Giddings and Eyring describes generally the chromatographic separation process and is therefore applicable also to the SEC separation (*13*). However, the probabilities k and k' in equation (9) are difficult to interpret directly in SEC because the mass transfer between MP and SP in SEC follows from the formation of an equilibrium which is continuously disturbed by the movement of MP. The irreversible process of attaining equilibrium is described by irreversible or non-equilibrium thermodynamics (*10*). The separation is a result of the mass transfer between the analyte fractions in the MP and in the accessible part of the pore volume of the SP (*31,32*) combined with longitudinal movement of the analyte in the MP. For the description of the separation in terms of this equilibrium a combinatorical approach based on division of the volume coordinate, V_0, into plates of size ΔV was developed (*12*). An elution curve is then obtained as the longitudinal concentration profile in the separation column develops in time and is observed in one place, namely in the concentration detector at the end of the column. This model describes the sequence in time of fractions of separated polymer occurring at a given point in space. Each fraction appearing consecutively in the detector is part of a different concentration profile. The sequence of the observed fractions is described by the negative binomial distribution (*12,33*) and its variance is expressed by (*12*)

$$\sigma^2 = V_0 \Delta V \frac{1-p}{p^2} \tag{12}$$

The following approximate values were found (12) $\Delta V \approx 1.81 \times 10^{-4}$ mL for toluene and $\Delta V \approx 3.18 \times 10^{-3}$ mL for a polystyrene standard of $M = 6.25 \times 10^5$ Da analyzed on two columns of length 23 cm filled with PL gel Mixed B, LS, resin of particle size 10 µm, separating molecular weights in the approximate range 400–10 000 000 Da, and employing a Showdex differential refractometer as a detector.

Using the relation between the variances expressed in units of volume (mL) and time (min), respectively, σ^2 and

$$\sigma^2_{min} = \sigma^2 / r_f^2 \tag{13}$$

where r_f^2 is flow-rate in mL.min^{-1}, the size of the equilibrium-displacement step can be expressed in units of time

$$\Delta t = \frac{\sigma^2_{min}}{t^\circ} \frac{p^2}{1-p} \tag{14}$$

which makes it possible to express constants k and k' from equation (9) as (13)

$$k = 2(1-p)/\Delta t \tag{15}$$

and

$$k' = kp/(1-p) \tag{16}$$

The relation between the kinetic and equilibrium description will be demonstrated with an example. Figure 1 shows a comparison of the elution curve of toluene in tetrahydrofuran as MP at $r_f = 0.5$ mL.min^{-1}, with the theoretical curve constructed according to equation (9) for $k = 438.3$ min^{-1} and $k' = 474.5$ min^{-1}, calculated for $p = 0.53$, $t_e = 38.9$ min, $y = 19.45$ mL, and $\sigma^2_{min} = 8.0 \times 10^{-2}$ min^2, according to equations (14)-(16).

Figure 1 also shows a graphical estimation of the standard deviation, $\sigma_{min} = 0.283$ min as 1/4 of the distance of the intersections of the tangents through the inflexion points of the elution curve with the baseline obtained on a system characterized by $t^\circ = 20.52$ min.

The accordance of curves in Figure 1 demonstrates that using the kinetic (k and k') and equilibrium (p) formalism in the description of the chromatographic separation is equivalent. Both descriptions of the separation process, however, are imperfect as they do not take into consideration the non-ideality effects discussed in the next section.

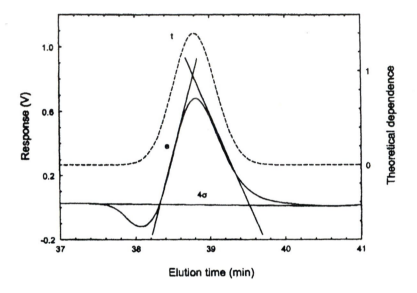

Figure 1. Comparison of the elution curve of toluene (solid curve), demonstrating the graphic determination of the standard deviation, σ_{min}, with theoretical curve (dashed) calculated according to equation (9).

Sources of Band Broadening

There are several sources of band broadening. The variance σ^2 is the sum of the contributions which are divided, according to the sources of broadening, into extracolumn and intracolumn (denoted by subscripts)

$$\sigma^2 = \sigma_{extra}^2 + \sigma_{intra}^2 \tag{17}$$

and the intracolumn contributions are divided according to whether they are related to the separation as discussed in the previous section (subscript 'sep'), or not (subscript 'nsep'), as

$$\sigma_{intra}^2 = \sigma_{sep}^2 + \sigma_{nsep}^2 \tag{18}$$

where σ_{nsep}^2 is composed of several sources of band-broadening, such as longitudinal diffusion (34), flow-tortuosity (35), and viscosity effects (12), all of which are difficult to separate from one another.

The contributions to σ^2 are frequently discussed using as background an equation frequently referred to as the van Deemter equation (36)

$$\sigma^2 = a + b/r_f + cr_f \qquad (19)$$

where the constants a, b and c are associated, in the first approximation, with eddy diffusion, longitudinal diffusion, and mass transfer, respectively (36).

Figure 2 presents the dependences of σ^2 on r_f found from SEC analyses of toluene and of a polystyrene standard of $M = 6.25 \times 10^5$ Da analyzed in tetrahydrofuran on the system described above (with DAWN-DSP-F, Wyatt Techologies multi-angle light scattering detector), together with curves calculated from equation (19) using values of a, b and c found by linear regression (12). For both analytes the minimum is approximately at $r_f = 0.5$ mL.min^{-1}. With increasing r_f, σ^2 rises due to insufficient mass transport of the analyte into pores; on the contrary, with decreasing r_f this term rises due to longitudinal diffusion.

It is interesting to compare the contribution of the broadness of the elution curve, expressed as variances, due to band broadening with that due to sample polydispersity, expressed as the $\overline{M}_w / \overline{M}_n$ ratio. A narrow MWD can be approximated by the log-normal function (equation (5)), which results in a theoretical Gaussian elution curve with standard deviation given by (37-39)

$$\sigma_{MWD}^2 = \frac{\ln \overline{M}_w / \overline{M}_n}{B^2} \qquad (20)$$

and the variance of the real elution curve is

$$\sigma_{EC} = \sqrt{\sigma_{MWD}^2 + \sigma^2} \qquad (21)$$

For a real separation system, values of σ close to 0.3 mL are found (22,40). Figure 3 shows a comparison of the elution curves having the standard deviation $\sigma = 0.3$ mL, σ_{MWD} and σ_{EC} for two samples of $\overline{M}_w / \overline{M}_n = 1.5$ and 1.01 calculated for a separation system characterized (22) by the calibration slope $B = 0.955$ mL^{-1}. For the former sample, the shape of the elution curve (σ_{EC}) is close to that of theoretical elution curve (σ_{MWD}) but for the latter the shape is close to that of spreading function (σ), which illustrates the fact that for narrow-MWD samples a symmetrical Gaussian elution curve is frequently obtained (22). On the other hand, with increasing concentration and/or M a non-ideal flow through the separation system occurs, caused by viscosity effects and hydrodynamic interaction of molecules, and the elution curve is asymmetric even if there is no interaction (penetration of molecules into the pores of the SP) of the analyte and no separation occurs.

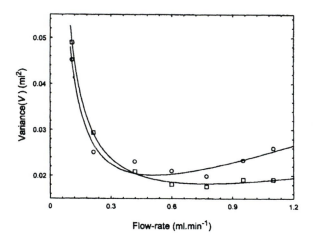

Figure 2. A comparison of the dependences on flow-rate of the variance of the spreading function, for toluene (○) and a polystyrene standard of M = 6.25 × 10⁵ Da (□). The curves were calculated according to equation (19) by linear regression. For details, see text. (Adapted with permission from reference (12). Copyright Elsevier.)

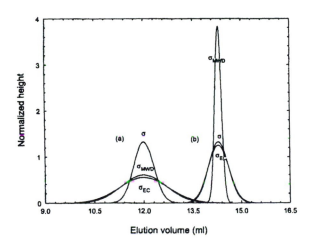

Figure 3. A comparison of Gaussian curves of standard deviation given by $\sigma = 0.3$ mL, σ_{MWD} and σ_{EC} values calculated, respectively from equations (20) and (21), for polymer samples of $\overline{M}_w / \overline{M}_n = 1.5$ (curves (a)) and 1.01 (curves (b)).

Band Broadening in SEC with Multiple Detection

The broadening of the elution curve of a mass-sensitive detector is described by equation (1). An analogoous equation holds for the signal of a molecular-weight-sensitive detector (23)

$$F(V)M_d^a(V) = \int_{-\infty}^{\infty} M^a(V)W(y)G(V,y)\,dy \qquad (22)$$

where $M_d(V)$ is the experimental molecular weight and a the exponent of the Mark-Houwink-Sakurada equation, $[\eta] = KM^a$ relating intrinsic viscosity, $[\eta]$, and M in viscometric detection; $a = 1$ in light-scattering detection.

Equation (22) can be solved numerically (23), i.e., the function $M^a(V)W(y)$ can be found using the same numerical deconvolution methods (14-26) as for solving equation (1), and the analytical solution (14) of equation (1) can be extended (38,41) to the solution of equation (22). From the solution of the two equations it follows that the local calibration dependence, i.e., the logarithmic dependence of M vs. V, is rotated due to band broadening and the absolute value of its slope decreases. This physically corresponds to mixing of contributions, as expressed by equation (1), of species of different M at each value of V, which is a process inverse to the separation resulting in decreasing the local weight-to-number-average molecular weight ratio, $(\overline{M}_w / \overline{M}_n)_d$. With increasing σ^2, i.e., with decreasing resolution, the number-average approaches to weight average molecular weight, $\overline{M}_n \to \overline{M}_w$ but the experimental value of $\overline{M}_w$ is unchanged (41). When the MWD of the sample can be approximated by a log-normal distribution (29) the linearity of local calibration is preserved (18,19,38) and can be expressed by (41)

$$\ln \overline{M}_w(V) = (1-\Sigma)\ln \overline{M}_w - \frac{\Delta}{2}\ln\sqrt{\overline{M}_w\overline{M}_z} - \frac{Z}{2} + \left(\Sigma + \frac{\Delta}{2}\right)(A+BV) \qquad (23)$$

where Σ was defined by equation (7),

$$\Delta = \frac{4\delta B}{2\sigma^2 B^2 + \beta^2} \qquad (24)$$

and

$$Z = \frac{2(\delta B)^2}{2\sigma^2 B^2 + \beta^2} \qquad (25)$$

where δ is the error in interdetector volume and will be discussed later. For the local weight-to-number-average molecular weight ratio, the solution gives a lower value (41)

$$(\overline{M}_w / \overline{M}_n)_d = (\overline{M}_w / \overline{M}_n)^{\Sigma+\Delta} \qquad (26)$$

On the other hand, the local $(\overline{M}_w/\overline{M}_n)_c$ ratio given by equation (8) is higher than the true one. The value of the true ratio, $\overline{M}_w/\overline{M}_n$, which is sought, is between those of $(\overline{M}_w/\overline{M}_n)_c$ and $(\overline{M}_w/\overline{M}_n)_d$. The true ratio can be obtained from these two values by solving equations (8) and (26) for $\overline{M}_w/\overline{M}_n$ by changing σ so that the values of $\overline{M}_w/\overline{M}_n$ obtained from the two equations coincide. This can be done numerically by a procedure based on equations (8) and (26) (22) or graphically by plotting (22)

$$\left(\overline{M}_w/\overline{M}_n\right)_{c,corr} = \left(\overline{M}_w/\overline{M}_n\right)_c \exp\left[-\sigma^2 B^2\right] \tag{27}$$

derived from equation (8) and denoting the values of $(\overline{M}_w/\overline{M}_n)_c$ corrected by particular values of σ as $(\overline{M}_w/\overline{M}_n)_{c,corr}$, where

$$\left(\overline{M}_w/\overline{M}_n\right)_{d,corr} = \left(\overline{M}_w/\overline{M}_n\right)_d^{1/\Sigma_{corr}} \tag{28}$$

is derived from equation (26) for $\delta = 0$ in equation (28), $(\overline{M}_w/\overline{M}_n)_{c,corr}$ denotes the values of $(\overline{M}_w/\overline{M}_n)_d$ corrected by a particular value of σ, Σ_{corr} is calculated using the correct slope, $B = -0.955$ mL^{-1}, of the calibration from equation (7) as

$$\Sigma_{corr} = \frac{\beta_{c,corr}^2/B^2}{2\sigma^2 + \beta_{c,corr}^2/B^2} \tag{29}$$

where $\beta_{c,corr}$ is

$$\beta_{c,corr}^2 = 2 \times \left(\ln\left(\overline{M}_w/\overline{M}_n\right)_c - \sigma^2 B^2\right) \tag{30}$$

The application of the graphical method is demonstrated in Figure 4 for the separation system characterized by $B = -0.955$ mL^{-1} and several examples of different $\overline{M}_w/\overline{M}_n$ ratio. The starting values of $(\overline{M}_w/\overline{M}_n)_d$ and $(\overline{M}_w/\overline{M}_n)_c$ were calculated from equations (8) and (26) for $\sigma = 0.3$ and the corrected values from equations (27) and (28) in dependence on σ. All curves intersect at $\sigma = 0.3$ mL at correct values of $\overline{M}_w/\overline{M}_n$ denoted at the intersection points of the curves.

The fact that a broadened elution curve results into an apparent MWD narrower than the true one can be explained on the basis of the relation between the elution curve, $F(V)$, and apparent MWD, denoted $f(\log M)$ (42)

$$f(\log M) = -B \times F(V) \tag{31}$$

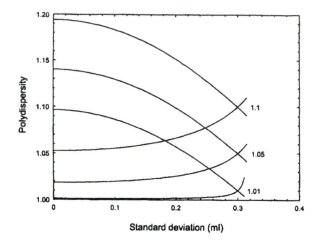

Figure 4. A comparison of curves calculated from equations (27) (descending curves) and (28) (ascending curves) for $\overline{M}_w / \overline{M}_n$ denoted at the intersection points of the curves.

The increasing absolute value of slope B results in a broader MWD and vice versa.

The method discussed above assumes that interdetector volume (IDV) is correctly determined ($\delta = 0$). The slope of the experimentally found local calibration is strongly influenced by the error δ in IDV determination. This error cannot be separated from the change of slope caused by band broadening, described by equation (23), unless the value of the IDV is determined by an independent method. The solution to this problem is discussed in the following section.

Interdetector Volume

From equations (23) and (26) it follows that the slope of the local calibration curve and the value of $(\overline{M}_w / \overline{M}_n)_d$ depend on the difference δ between the correct value of the IDV and the value used in the calculations. This error becomes important for narrow-MWD samples and compensates for the effect of band broadening. When analyzing one sample these two effects cannot be separated from each other unless there is an independent method of IDV determination. This method is based on s-detection, i.e., on the equivalence between the root-mean-square radius (radius of gyration), $\langle s^2 \rangle^{1/2}$, and M, expressed by equation

$$s = kM^{\alpha} \tag{32}$$

In solutions of polydisperse samples, $\langle s^2 \rangle^{1/2}$ correlates with the s-average molecular weight (43)

$$\overline{M}_s = \left[\overline{M}_w^{-1} \sum_i w_i M_i^{1+\alpha} \right]_{s-}^{1/\alpha} \tag{33}$$

For theta conditions, $\alpha = 1$ and the $\overline{M}_s = \overline{M}_z$. In thermodynamically good solvents, the polymer coil expands (44) and $\alpha > 1$. From the theoretical analysis it follows that the differences between local values of $\overline{M}_s(V)$ and $\overline{M}_w$ found by the dual detection (LS-detection) are negligible (45).

The local calibrations calculated for several values of interdetector volume (IDV) for a polystyrene reference standard of $M = 1.6 \times 10^6$ Da together with the local calibration obtained by the s-detection, are in Figure 5. A high sensitivity of the local calibration to small changes of the IDV value (denoted with the local calibrations) shows that IDV cannot be found reliably from geometric considerations (46) and the s-detection makes it possible to find its effective value (45).

Local Polydispersity

Band broadening is a source of local non-uniformity in M (polydispersity), i.e., non-uniformity in an infinitesimal volume of the eluent. For the log-normal model function, it can be expressed by (41)

$$\left(\overline{M}_w / \overline{M}_n \right)_1 = \left(\overline{M}_w / \overline{M}_n \right)^{1-\Sigma} \tag{34}$$

For $\sigma \to 0$, $\left(\overline{M}_w / \overline{M}_n \right)_1 \to 1$ for $\sigma > 0$, this ratio rises with increasing σ up to $\left(\overline{M}_w / \overline{M}_n \right)_1 \to \overline{M}_w / \overline{M}_n$.

The local polydispersity caused by band broadening should not be confused with non-uniformity due to due to long-chain branching (47). The latter comes from differences in hydrodynamic volume of species eluting at a given V (48,49).

316

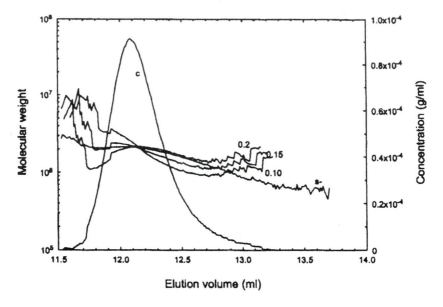

Figure 5. Comparison of the local calibrations obtained for different values of interdetector volume used in the software calculations with the local calibration obtained independently by the s-detection based on equation (32), relating the radius of gyration and molecular weight (labeled on the curves).

Acknowledgments

The author gratefully acknowledges helpful discussions of J. Kahovec, the loan of the DAWN laser photometer from Wyatt Technologies Ltd. as well as support by the Grant Agency of the Czech Republic (project No. K4055109).

References

1. Grubisic, Z.; Rempp, P.; Benoît, H. *J. Polym. Sci., Part B: Polym. Phys.* **1967,** 5, 753-759.
2. Moore, J.C. *J. Polym. Sci., Part A: Polym. Chem.* **1964,** 2, 835-843.
3. Grubisic-Gallot, Z.; Picot, M.; Gramain, Ph.; Benoît, H. *J. Appl. Polym. Sci.* **1972,** 16, 2931-2945.
4. Haney, M.A. *J. Appl. Polym. Sci.* **1985,** 30, 3023-3036.
5. Haney, M.A. *J. Appl. Polym. Sci.* **1985,** 30, 3037-3049.

6. Ouano, A.C. *J. Polym. Sci., Part A-1: Polym. Chem.* **1972**, 10, 2169-2180.
7. Ouano, A.C. *J. Chromatogr.* **1976**, 118, 303-312.
8. Wyatt, P.J. *Anal. Chim. Acta* **1993**, 272, 1-40.
9. Berek, D. *Mater. Res. Innovations* **2001**, 4, 365-374.
10. Giddings, J.C. Unified separation science. J. Wiley Interscience; 1991; p 38.
11. Giddigns, J.C.; Eyring; H. *J. Chem. Phys.* **1955**, 59, 416-421.
12. Netopilík, M. *J. Chromatogr. A* **2002**, 978, 109-117.
13. Netopilík, M., *J. Chromatogr. A,* in press.
14. Tung, L.H. *J. Appl. Polym. Sci.* **1966**, 10, 375-385.
15. Procházka, O.; Kratochvíl, P. *J. Appl. Polym. Sci.* **1986**, 31, 919-928.
16. Busnel, J.P.; Foucault, J.P.; Denis, F.; Lee, W.; Chang, T. *J. Chromatogr. A,* **2001** 930, 61-71.
17. Tagaki, T. *J. Chromatogr.* **1990**, 506, 409-416.
18. Lederer, K.; Imrich-Schwarz, G.; Dunky, M. *J. Appl. Polym. Sci.* **1986**, 32, 4751-4760.
19. Jackson, C. *J. Chromatogr.* **1993**, 645, 209-217.
20. Jackson, C. *Polymer* **1999**, 40, 3735-3742.
21. Vega, J.R.; Meira, G.R. *J. Liq. Chromatogr. Relat. Technol.* **1993**, 645, 901-919.
22. Netopilík, M.; Podzimek, S.; Kratochvíl, P. *J. Chromatogr. A* **2001**, 922, 25-36.
23. Netopilík, M. *Polym. Bull.* **1982**, 77 575-582.
24. Pierce, P.E. and Armonas, J.J., *J. Polym. Sci., Part C* **1968**, 1, 23-29.
25. Ishige, T.; Lee, S.-I.; Hamielec, A.E. *J. Appl. Polym. Sci.* **1971**, 15, 1607-1622.
26. Rosen, E.M.; Provder, T. *J. Appl. Polym. Sci.* **1971**, 15, 1687-1702.
27. Berger, K.C. *Makromol. Chem.,* 179 (1978) 719-732.
28. Wesslau, H. *Makromol. Chem.,* **1956**, 20, 111-142.
29. Kotliar, A.M. *J. Polym. Sci., Part A* **1964**, 2, 4303-4325.
30. Hamielec, A. *J. Appl. Polym. Sci.* **1970**, 14, 1519-1529.
31. Yau, W.W., Kirkland, J.J., Bly, D.D. Modern size-exclusion liquid chromatography, J. Wiley and Sons: New York, 1979.
32. Kirkland, J.J. (Ed.) Modern practice of liquid chromatography, Wiley Interscience: New York, 1971, p 237.
33. Abramowitz, M.; Stegun, I. Handbook of mathematical functions; Dover Publ. Ltd., New York, 1965.
34. Striegel, A.M. *J. Chromatogr. A* **2001**, 932, 21-31.
35. Knox, J.H. *J. Chromatogr. A* **1999**, 831, 3-15.
36. van Deemter, J.J.; Zuiderweg, F.J.; Klinkenberg, A. *Chem. Eng. Sci.* **1956**, 5, 271-289.
37. Kendrick, T.C. *J. Polym. Sci. A2* **1969**, 7, 297-307.
38. Netopilík, M. *Polym. Bull.* **1983**, 10, 478-481.

39. Aust, N.; Parth, M.; Lederer, K. *Int. J. Polym. Anal. Charact.* 6, **2001,** 245-260.
40. Cheung, P.; Lew, R.; Balke S.T.; Mourey, T.H. *J. Appl. Polym. Sci.* **1993,** 47, 1701-1706.
41. Netopilík, M. *J. Chromatogr. A* **1998,** 793, 21-30;.
42. Berger, H.L.; Schulz, A.R. *J. Polym. Sci. A* **1965,** 319, 3643-3648.
43. Kratochvíl, P. Classical light scattering from polymer solutions (Polymer Science Library 5, Ed.: A.D. Jenkins), Elsevier: Amsterdam, 1987.
44. Flory, P.J. Principles of polymer chemistry, Cornell University Press, Ithaca: New York, 1953.
45. Netopilík, M.; Podzimek, S.; Kratochvíl, P. *J. Chromatogr. A,* accepted for publication
46. Bruessau, R., in Liquid Chromatography of Polymers and Related Materials II, J. Cazes, ed., Marcel Dekker, Inc., New York, 1980, pp. 73-93.
47. Hamielec, A.; *Pure Appl. Chem.,* **1982,** 54, 293-307.
48. Park, W.S.; Graessley, W.W. *J. Polym. Sci., Polym. Phys. Ed.* **1977,** 15, 85-95.
49. Netopilík, M.; Kubín, M.; Schultz, G.; Vohlídal, J.; Kössler, I.; Kratochvíl, P. *J. Appl. Polym. Sci.* **1990,** 40, 1115-1130.
50. Netopilík, M. *J. Chromatogr. A,* **1998,** 809, 1-11.

Indexes

Author Index

Subject Index

A

Acrylonitrile. *See* Poly(styrene-*co*-acrylonitrile) (SAN)

Aggregation
  sodium polystyrene sulfonate (PSS), 46–47
  See also Protein–polymer conjugates

Aluminum speciation, forest soil, 172

Alzheimer's disease, trace elements in brain by SEC–inductively coupled plasma mass spectrometry, 178–179

Amniotic liquid, Pb-bound ligand in, by SEC–inductively coupled plasma mass spectrometry, 179

Analytical temperature rising elution fractionation (ATREF)
  assessing heterogeneity in polyolefins, 223, 225
  ethylene 1-butene and 1-hexene copolymers, 227*f*

Angles, scattering measurements, 27–28

Anion-exchange chromatography (AEC), oligosaccharides, 7

Architecture
  branching in polyolefins, 62–65
  conformation, 3
  Mark–Houwink relation, 53–54
  measurements and end-use effects, 5*t*
  particle form factor for spheres, random coils, and rods, 55*f*
  radius of gyration, 55–56
  ratio R_g/R_H for polymer architectures, 67*t*
  semi-rigid polymers, 71–73
  star polymers, 67–71

Atmospheric pressure chemical ionization (APCI)
  electrospray ionization (ESI) as preferred method, 203–204
  mass spectrometers handling interfaces, 203
  on-line analysis of Triton X-100, 204*f*

Automatic Continuous Mixing (ACM)
  batch experiment, 18, 19*f*
  poly(vinylpyrrolidone) (PVP) with ACM, 40, 41*f*
  two-component polymer solutions, 40, 41*f*

B

Band broadening
  comparing local calibrations, 316*f*
  formulation, 304–308
  Gaussian elution curve, 309, 311*f*
  interdetector volumes, 314–315
  local calibration dependence, 304
  local polydispersity, 315
  multiple-detection size exclusion chromatography, 312–314
  polymer chromatography, 290–291
  single detector, 303
  sources, 308–309
  See also Size exclusion chromatography (SEC)

Barley
  size exclusion chromatography/multiple angle laser light scattering (SEC/MALS), 145*f*
  See also Glucan polymers, soluble

Base pair sequence, measurements and end-use effects, 5*t*

Berry fit method

323

DATE DUE